# SALUMI

Also by Michael Ruhlman and Brian Polcyn

*Charcuterie*

OTHER FOOD BOOKS BY MICHAEL RUHLMAN

*The Making of a Chef: Mastering Heat at the Culinary Institute of America*

*The French Laundry Cookbook* (with Thomas Keller and Susie Heller)

*The Soul of a Chef: The Pursuit of Perfection*

*A Return to Cooking* (with Eric Ripert)

*Bouchon* (with Thomas Keller, Jeffrey Cerciello, and Susie Heller)

*The Reach of a Chef: Professional Cooking in the Age of Celebrity*

*Under Pressure: Cooking Sous Vide* (with Thomas Keller, Susie Heller, Amy Vogler, Jonathan Benno, and Cory Lee)

*The Elements of Cooking: Translating the Chef's Craft for Every Kitchen*

*Ratio: The Simple Codes Behind the Craft of Everyday Cooking*

*Live to Cook* (with Michael Symon)

*Ad Hoc at Home* (with Thomas Keller, Dave Cruz, Susie Heller, and Amy Vogler)

*Ruhlman's Twenty: 20 Techniques, 100 Recipes, A Cook's Manifesto*

# SALUMI

*The Craft of Italian Dry Curing*

MICHAEL RUHLMAN AND BRIAN POLCYN

Photographs by Gentl & Hyers/Edge

Styling by Andrea Gentl

Drawings by Alan Witschonke

W. W. Norton & Company

New York • London

Printed in the United States of America
First Edition



For information about special discounts for bulk purchases, please contact W. W. Norton Special Sales at specialsales@wwnorton.com or 800-233-4830

Manufacturing by Courier Kendallville
Book design by Jan Derevjanik
Production manager: Devon Zahn

Library of Congress Cataloging-in-Publication Data

Ruhlman, Michael, 1963–
Salumi : the craft of Italian dry curing / Michael Ruhlman and Brian Polcyn ; drawings by Alan Witschonke Illustration ; photography by Gentl & Hyers. — 1st ed.
p. cm.
Includes bibliographical references and index.
ISBN 978-0-393-06859-7 (hardcover)
1. Dried meat—Italy. I. Polcyn, Brian. II. Title.
TX609.R844 2012
641.4'4—dc23

2012008184

W. W. Norton & Company, Inc.
500 Fifth Avenue, New York, N.Y. 10110
www.wwnorton.com

W. W. Norton & Company Ltd.
Castle House, 75/76 Wells Street, London W1T 3QT

1 2 3 4 5 6 7 8 9 0

*To our mothers, Josephine Polcyn and Carole Ruhlman,*
*who taught us, by example, the love of cooking.*

# Contents

# 1. Getting Started

# Making the Ancient New

Something truly amazing has been underway in America for more than a decade now. The issue is so vast that for decades we lost sight of it, maybe even never saw it in this country until recently: *the importance of food*. If you'd tried to argue that food was not "important" to one of our early ancestors, or to someone today who doesn't have enough of it, they'd look at you as if you'd been living on another planet. But because food became so easily attained in the developed world, thanks to shipping, refrigeration, and infinite-shelf-life processing, we took it for granted. And we have only recently become aware of its importance on a national scale because our food supply has become imperiled and food-related illnesses, from bacterial contamination to diabetes, have begun to make us sick on an epidemic scale. The ramifications of this relatively new awareness are diverse: the FDA debates regulating how much salt companies can put in processed foods, physicians argue about whether or not food is as chemically addictive as alcohol or nicotine and why children's food allergies have become as common as colds, and Congress debates farms subsidies. We've turned chefs, once anonymous tradespeople, into celebrities. Food issues and cooking as sport have become common entertainment, food bloggers are attracting six-figure book deals, and farmers' markets are flourishing throughout the country.

Amid this *sturm und drang*, a few truly wonderful changes have moved in like soothing waves through our culture. Changes so fine and unlikely that Brian and I believe that there has never been a better time in this country's history to be a cook and to take pleasure in the cooking and sharing and eating of food.

America has always been a culture that embraces the new. But in the case of our food post World War II, "new" was not good for us. New was in fact bad in a lot of ways, and we have only in the past decade begun to recognize it. The trans fats in margarine, a "healthy" alternative to butter, actually made it unhealthy. We learned that there was high-fructose corn syrup in our bread and that the dyes added to processed food to make it more appealing were harming our kids. The antibiotics used to keep our cows healthy created a new bacterium that has killed and maimed.

But thanks to a few voices in the food world, "old food" and "slow food" have become new. And America has embraced this. We can only hope that, as the newness wears to inevitable age, we still sense the pleasure in the weathered surface and the clean, simple food that looks on the plate as it did coming out of the ground, that we recognize the power and importance of a well-made cheese, or a braised beef brisket, or potatoes mashed with whole milk, butter, and salt.

As ever, chefs have led the way in our new understanding, and their work and knowledge has filtered down into home kitchens. Alice Waters fostered a recognition of the value of naturally raised food we grow ourselves or that is grown by farmers near where we live when she opened Chez Panisse in 1971. Larry Forgione asked us to pay attention to our regional cuisines when he opened An American Place in 1983, celebrating a country so huge that what we grow or catch in one corner is in the other corner vastly different: grapefruit and grouper in the southeast, apples and salmon in the northwest, and sour cherries and walleye in between.

In Italy, Carlo Petrini, a writer and eco-provocateur, appalled by the proliferation of American fast food outlets in Rome, spearheaded a food movement called Slow Food in 1986. It has spread its promotion of naturally and sustainably raised and harvested food worldwide.

In the 1990s, more and more chefs began demanding excellence in their products and found farmers and foragers willing to work with them to achieve it. And soon that same search for excellence filtered down to everyone who liked to cook, and supermarkets worked to satisfy their customers with once unheard-of ingredients: morel mushrooms, habanero peppers, and ostrich eggs in the grocery store.

It was these changes that allowed us to publish *Charcuterie: The Craft of Salting, Smoking, and Curing* in 2005 with uncommon success. Yet the unlikeliness of that has to be underscored. *Charcuterie* is a book devoted to the French tradition of preserving meats by curing and confiting them, with recipes whose two principal ingredients are fat and salt. *Fat and salt:* villains number one and number two on

American nutritionists' Most Wanted list. At the same time, America had become obsessed with the fast and easy meal, and even 30-minute meals took too long. We also became terrified of germs and bacteria, getting rid of perfectly good wood cutting boards and buying up all manner of anti-bacterial soaps, dumping everything in the fridge if the power goes out for more than four hours. Into this culture, we brought a book not just devoted to animal fat and salt, but also reflecting a full-on love affair with them, a sweaty, torrid embrace of them. Moreover, many of the recipes take not 30 minutes, but rather days, sometimes even months, to prepare—and recipes that ask you to *add* bacteria to your food, while telling you that if you don't do it right, it can kill you.

And chefs and cooks far and wide, bless them, embraced the book. Hundreds of readers—and bloggers—took up the call to cure their own bacon and confit their own duck. This is a food culture Brian and I are very glad to be a part of. And it encouraged us to continue our exploration of the powers of salt and the majesty of the hog in *salumi*, the Italian version of the French craft we came to adore—the slowest food of all. But salumi and charcuterie are not the same craft with different names. Salumi is a narrower, more focused, and more difficult craft, one that should be approached the way one might hunt wild boar: with knowledge, respect, the proper tools, and the recognition that you might have a good day and you might not, you might catch something and you might come up empty-handed—and that is part of the thrill of it. Because when you find that boar, and you will if you are willing to work for it, you can turn it into *cinghiale* sausages, the dry-cured wonders found throughout Italy, and the result is as thrilling as magic.

Nature is the greatest artist, we are not the first to say, and this is what salumi is really about: taking what nature gives us and doing as little as possible to it to make it the best it can be.

Over the past decade, dozens of salumerias and charcuteries have opened throughout the country. Marc Buzzio, whose ancestors hailed from Biella in the Piedmont region of northern Italy, has kept the family business up and running by curing sausages in New York City the way his dad and granddad had for decades.

America has a long tradition of smoked country ham, yet that all but vanished in the wake of factory farming and factory production. With few exceptions, like Buzzio's family and Nancy Newsom's family, who has been curing small batches of meats—in this case superlative country hams—in Kentucky since 1917, not only was great salumi unavailable, most people didn't even know what decent salami and dry-cured ham were. In the span of a decade, America went from a country that knew only cooked salami from Oscar Meyer and Boar's Head, factory salami, to one where *salumi* every bit as good as the finest in Italy is available to all from numerous sources.

## Let's Not Get Confused Here: Important Definitions

- *Salumi* is the Italian word for salted and cured meats. (*Salume* is the singular form of the word; we almost always use the plural.) Salumi include pancetta, prosciutto, coppa, and salami.

- Salami are dry-cured sausages. If you have only one of them, technically it's called a salame.

Armandino Batali, chef Mario's dad, opened Salumi Artisan Cured Meats in Seattle in 1999. In 2001, Herb and Kathy Eckhouse started La Quercia in Norwalk, Iowa, to dry cure hams that are considered by experts to be every bit as good as the best prosciutto in Parma. In 2006, chef Paul Bertolli opened Fra' Mani in Oakland, California, where he makes handcrafted salumi for retail and online sales. And, in the same city, chef Chris Cosentino started Boccalone in 2007. That same year, Cristiano Creminelli brought his family's salumi business, also in the area of Biella, to the United States, opening Creminelli Fine Meats in Salt Lake City, Utah. Dozens of younger chefs, fascinated by the ancient craft, have since opened salumi stores throughout the country, and restaurant chefs have set up curing environments in their kitchens so that they can offer their own house-cured meats.

The ancient craft of salumi, which was scarcely known in this country, is now growing; the seeds have been planted and now have strong roots. As appreciation and demand for salumi grows, the craft has begun to thrive.

In this book, we explain and explore it, one of the oldest forms of cooking still practiced. We describe how it works, and the tools required to make it happen, and we provide the recipes and techniques to make the whole range of preparations called salumi. This is a craft that requires time, attention, care, and not a little ingenuity. Sometimes you may fail, but when you succeed, the rewards are more than commensurate with the effort. Indeed, when you eat great salumi, it can seem like magic. Because in a way it is.

## Italian Salumi Versus American Salumi Is Not the Issue: Some Perspective

Is authentic Italian salumi possible in America? No, nor would we wish it to be. But can we make extraordinary *American* salumi? Yes, and artisans and cooks make it every day throughout the country.

The distinction between the two is an important one. Italian salumi is the result of a specific landscape and an atmosphere providing a consistent range of temperatures and humidity fundamental to the creation of exquisite dry-cured meats. The breeds of hog differ from ours, as does what they eat, and this in turn affects the flavor of the finished salumi (see Hog Breeds, page 27). The same is true of all the animals we eat, but the effects of environment, species, and diet are especially significant in dry curing, which is a way of "cooking" meat without ever bringing it above room temperature.

Yet, as with wine, Americans have proven to be fast on the uptake, aggressive learners, as the products of the salumi makers mentioned above attest. Italy is where it all flourished, however, and it is from Italy that we have taken our inspiration and education.

Etruscans and Romans put their environment—an atmosphere favorable to dry curing, and *terroir* that favors the hog, the most valuable and extraordinary animal we eat—to good use to preserve their food. And the care of their salumi was a matter of survival. As a present-day salumi maker put it to us when we traveled to the mountain town of Colonnata, "Sixty years ago, there were no roads up here. We lived off the hog." His statement does not have the gravity on the page that it did on hearing it after we'd navigated a dozen impossibly narrow hairpin turns a thousand feet up into the Apuan Alps, the sheer white faces of the famous marble, carved out of the earth, shining in the sun along our way.

The people of Colonnata, who have mined marble there since before Michelangelo worked his own art with it, have always had to rely on the hog for survival. They preserved all cuts, surely, but they discovered that when the back fat (*lardo*) was cured in boxes made from the marble the town was built on, it was uncommonly delicious, far better than when simply cured in salt and aged in a cellar, as so much of the hog typically was. The hog fat was so delicate it melted on the tongue. A fine layer of crystalline salt coated its exterior, and when it was sliced thin, it retained a crisp line of salt to season and contrast the sweet soft fat.

Their *lardo di Colonnata* is the very definition of a regional specialty. It uses a material, marble, unique to the area, which gives the fat from hogs grown in this area qualities that cannot be developed in any other place. This is how local specialties happen. And sometimes local specialties become so popular that other marketers try to feed off the name. One lardo curer there told us that 200,000 kilograms of lardo are cured in those hills each year, but 9 million kilograms are sold with the Colonnata name. That is why the DOP stamp was created—DOP (*Denominazione di Origine Protetta*) indicates that the origin of the product is protected and that specific standards are maintained and overseen by the government.

Curing the hog in these isolated mountains began out of necessity, but the tradition has endured because the result is delicious. And because it's been made here for thousands of years, the salumi makers have gotten really, really good at it.

*Culatello* is another example of an extraordinary local specialty. It is the top and bottom round of the ham, the back leg of the hog, which is salted, sewn up in a hog bladder, tied, and dried for at least two or up to nearly four years. It's made in and around the city of Zibello, on the Po River in Emilia-Romagna. This area of Italy is probably best known for Parma ham, prosciutto, the dry-cured entire ham. The makers of culatello say that the air along their stretch of the Po River is too moist for whole hams to dry properly. Early settlers of the region took off the main muscles of the ham to make the large culatello and, from the other side of the ham, the cut called *fiocco*. Both cuts dried beautifully in the moist cellars along the Po. They hang for so long that they take on a flavor and texture that is unique to this small spot of earth.

However, the culatello tradition nearly died out, say its current producers. But, led by two brothers, Massimo and Luciano Spigaroli, farmers who had bought an ancient crumbling estate to cure culatello, the makers of this local specialty lobbied for

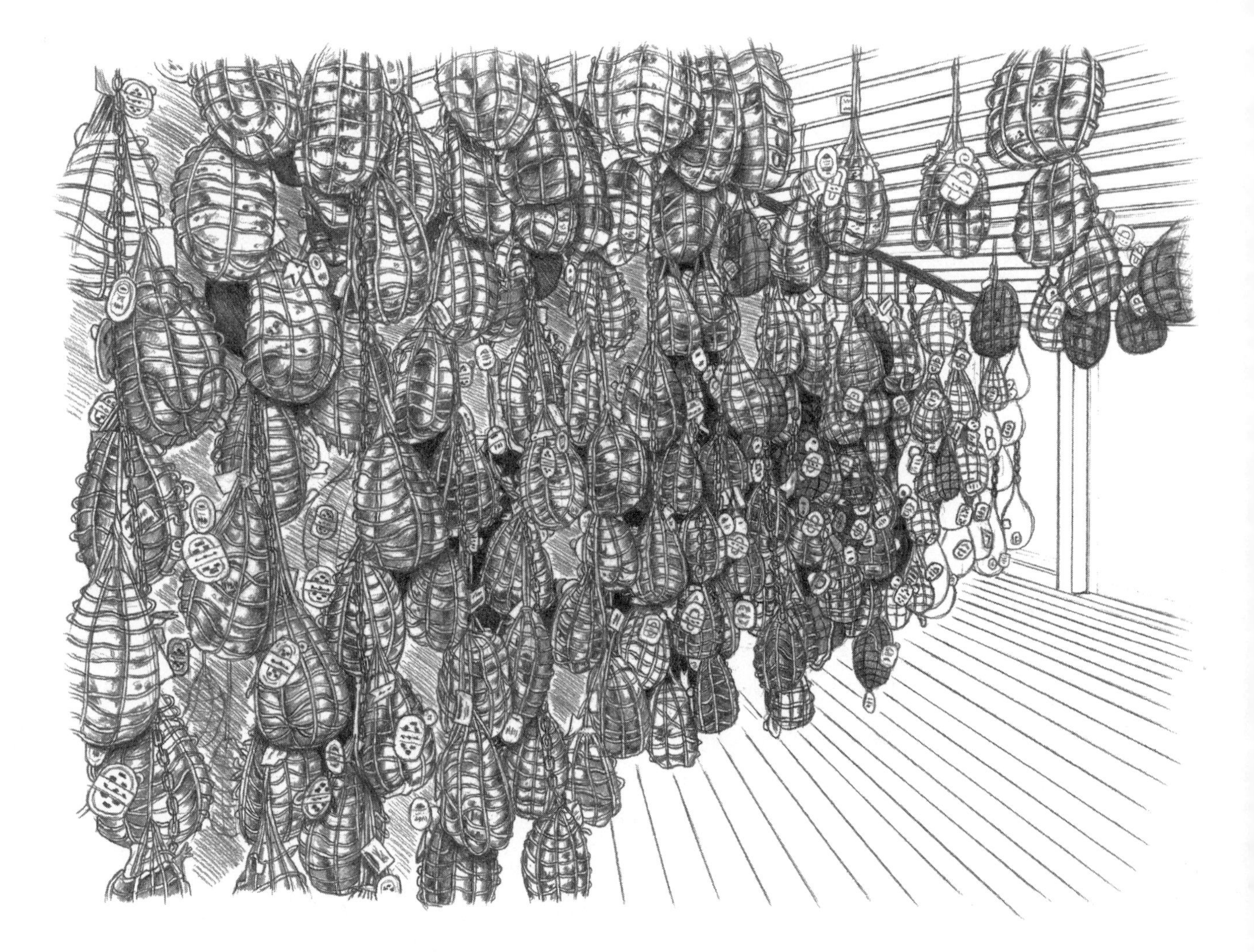

Massimo Spigaroli is the king of culatello. The cellar of his restaurant and inn, Antica Corte Pallavicina, beside the Po River in the province of Parma, in the region of Emilia Romagna, is filled with these prized boneless hams, some of which have been aged for as long as 45 months.

and got a DOP designation for their product, *culatello di Zibello*. Culatello is unique because of the hogs that thrive here—the Large Whites, the Nera Parmigiana, and the Mora Romagnola—and the humid storage areas at or below the river's level. (The Spigarolis' cellars are about 85 percent humidity). Thanks to the Spigarolis (for not only their exquisite product but also their savvy marketing skill), the culatello is alive and well, worth seeking out if you happen to be traveling along the Po River. It is, in fact, the best form of prosciutto we have ever tasted.

So, the craft of salumi is alive and well in most regions of Italy, but, with only a handful of exceptions (such as the Spigarolis), small producers now use electricity to control humidity and temperature. That control results in a consistent product, as previously salumi makers had to use their senses and experience (knowing when to open or close the windows to control humidity, for instance) to a greater degree. Unfortunately, electronic control has also ushered in an age of large commercial interests steamrolling myriad small producers out of business with the cheap prices that come with volume and speed. Yes, this happens everywhere: it's not just McDonald's in Italy, Italian producers do it as well—that's business.

It's clear to the people who create such food as *culatello di Zibello* and *lardo di Colonnata,* that the work and financial risk involved are worth it. But only some people are cut out to do the work. America is not the only country vulnerable to business bottom lines. Concern for what was happening to food in Italy, aided by entrenched sensibilities regarding food there and a fierce territorial pride in its products that pits one region against another, resulted in the rise of the Slow Food movement.

Salumi is an example of Slow Food and a testament to its importance.

Let's examine, by way of comparison, the food America has for nearly the past century worked so hard to put out. We don't have a food culture that developed over millennia. What food culture did arise was based on local subcultures created by huge waves of immigrants.

Food cultures evolve out of necessity. Throughout history, the bottom line has been to protect family and community, making sure all have enough to eat. Food culture has always been about making sure first that your family had enough to eat and, then, that those in your community upon whom you relied, for work, for commerce, for protection, also had enough to eat. Societies ate what was native to the area, because it was easiest; life was hard, and you made what food you could grow, gather, or catch taste as good as you could with whatever resources were at hand. And this became each group's food culture, its heritage, and it endured long after that society needed it for survival—the salumi of Italy, for instance, the confits of France, the smoked hams of Germany, and the preserved fish of both tropical and northern regions.

Resisting the forces of big business and globalization is hard, especially if you have so little to protect, so when frozen food and processed food began to pour into American grocery stores after World War II, there was little thought of losing anything in particular; there was, instead, an embrace of the new and an acquiescence to the advertising that promised ease and convenience.

But salumi is one example of old food that America is now embracing as new, and that we hope endures long enough to become old in this country. Indeed, among the main motivations of this book are, first, to ensure that this form of food preservation never dies and, second, that Americans come to understand both the historical importance of the craft of salumi as well as the pleasures of eating it.

Salumi is easy to do on a general level: throw some salt on some meat and let it dry, slice thin, and eat. To do it with excellence, however, takes an elusive combination of *terroir*, passion, and luck.

This book is about salumi, the Italian technique of dry curing and preserving meat, but it is also a continuation of our exploration of meat curing and sausage making in general. And although it's about curing different cuts of meat from different animals, primarily we embarked on this project to deepen our understanding of the hog, the culinary miracle upon which so much of the world has survived, and to deepen our knowledge of this ancient craft of salting and drying meat, once practiced as a matter of survival, today practiced for the unparalleled pleasures it provides, a reminder of our deep past, and a reflection of our humanity.

## The Big Eight: Simplifying the Confusing Terminology of Salumi

1. **Guanciale** • jowl
2. **Coppa** • neck/shoulder/loin
3. **Spalla** • shoulder
4. **Lardo** • back fat
5. **Lonza** • loin

5a. • tenderloin

6. **Pancetta** • belly
7. **Prosciutto** • ham, back leg
8. **Salami** • products made from ground or cut pieces of pork

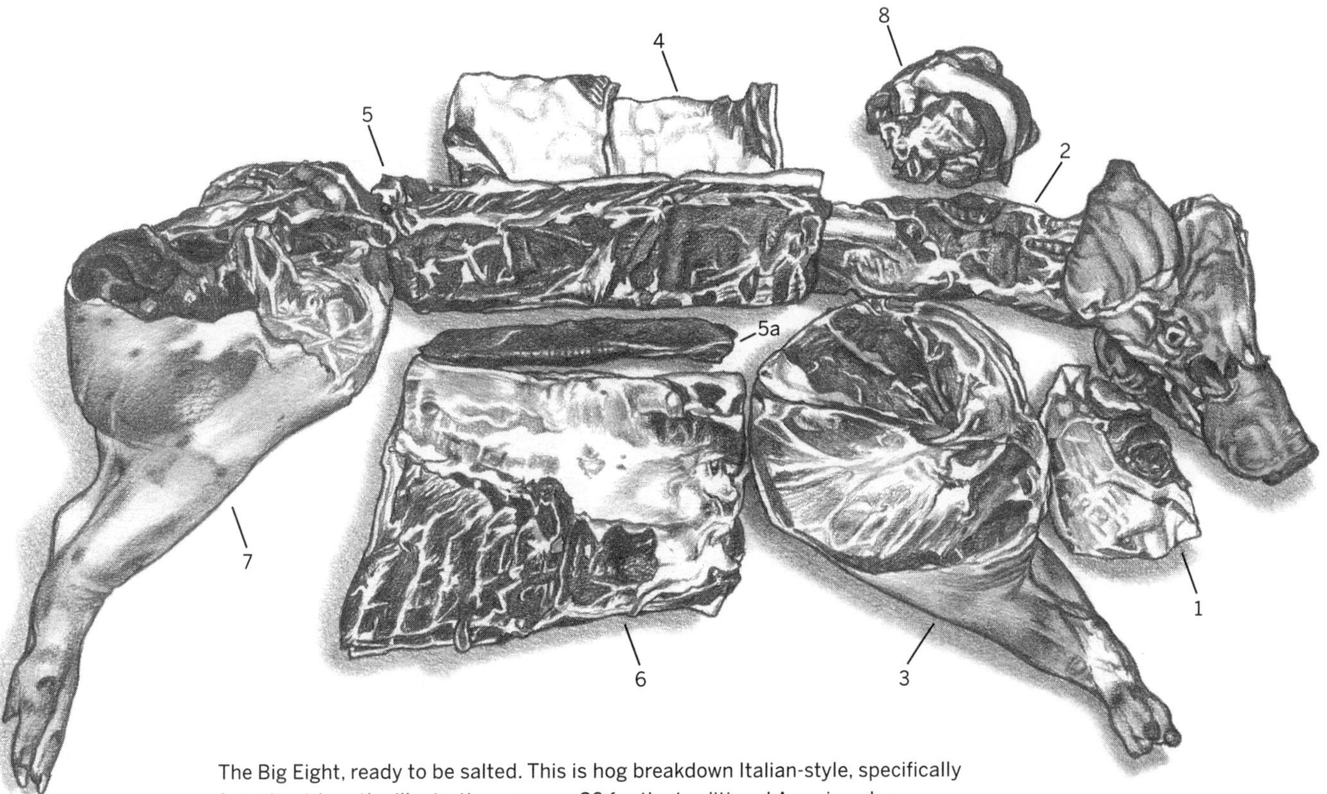


The Big Eight, ready to be salted. This is hog breakdown Italian-style, specifically for salumi (see the illustration on page 32 for the traditional American hog breakdown).

Walk into a salumeria almost anywhere in Italy and you will be confronted by a panorama of cured and cooked meats and sausages. Our first stop on a recent trip was Mosca, a high-end salumeria and specialty store in the city of Biella in the Piedmont region. We could choose from its long and glorious case any of the following sausages, most of which are dried: *salame crudo filzetta, piccoli storti dolcissimi, asinelli, bocconcini pura coscia, cacciatorino Napoli piccante, il grissino suino, salame mantovano tipico, salame rustico, salame da cuocere, salame d' la duja*, and *salame cotto di vitello fassone alla monferrina*, to name a few. And this before we noticed the ham section, the loin section, and the section offering no fewer than a dozen preparations of fatback and belly.

The variety is dizzying. It's like walking into a pork version of Willy Wonka's chocolate factory.

Travel south into Tuscany, and a salumeria of the same caliber will have the same products, but the names will differ, as they do next door in Umbria or way down south in Calabria. There are, in fact, dozens of dialects spoken throughout Italy—some Italians don't even speak Italian, just a dialect. Salumi terminology is subject to this same fragmentation. Trying to understand Italian salumi is like trying to understand Italian in a room full of dialect speakers.

Before we set out to explore Italian salumi, we would walk into a salumeria and gaze at the innumerable Italian names for the dried sausages with little clue as to what distinguished one from another beyond "hot" or "sweet" or "rustic." Further exploration only increased our confusion. For years, we've known soppressata as a large dry-cured salame made with any number of seasonings. One source said there were seven hundred or more published recipes for it. But walk into Sergio Falaschi's fine Tuscan salumeria and point to the enormous cylinder of dark cooked meat, three feet long and seven or eight inches in diameter, hanging in front of the shelves of dry-curing pork loin, and you will be told that it is soppressata. And the slices in the deli case show a cross section of chunks of meat, tongue, and jowl skin, cooked until tender, compressed into shape with a cloth casing and held fast by lots of gelatin—what I had always known as headcheese. Sergio's wife, Andrea, will nod and say, "*Si, soppressata.*"

We saw it called *coppa di testa* in another region. Two regions over, in Le Marche, soppressata is very high in fat and spreadable; in the south, it is a sliceable salame.

In the aforementioned Mosca in Biella, beside the salami, you will see traditional coppa, a neck-shoulder muscle that, with a high ratio of lean meat to fat, is particularly suited to dry curing. Walk into a salumeria in Umbria, in central Italy, and you will see that same muscle, with its characteristic cross sections of fat, called *capocollo*. Head farther south, and you will find that it's referred to as *filetto*.

Bring this up with Emanuelle Sbraletta at his Enotecha Salumeria in the ancient city of Bevagna, in Umbria, and he will admit it is confusing even for Italians.

To confirm, I said, "In the north, coppa, in Tuscany, and here in Umbria, capocollo?"

"No," he said. "In Tuscany, yes; here, no."

Again confused, I said, "But there is capocollo here."

"No, there is not."

I looked to Brian, who was equally confused.

"I show you," Emanuelle said, and called to the man behind the counter. "Papa, cut me some capocollo."

"But you just said you didn't have capocollo here."

Emanuelle closed his eyes and sighed, held his palms together, nodded, and said solemnly, "It is confusing even for us."

It needn't be confusing. In fact, in general overview, salumi couldn't be more simple or clear. While there are countless variations—dry-cured whole; ground and dry cured; ground and cooked; cooked whole; dry cured, then cooked—it is, as ever, the humans who complicate matters; the hog is simplicity and generosity itself.

The truth is this: there are eight basic products in Italy's pork salumi repertoire, and we explore them on pages 87–121: guanciale (jowl), coppa (neck), spalla (shoulder), lardo (back fat), lonza (loin), pancetta (belly), prosciutto (ham), and salami (ground meat). The vast array of Italian salumi can be easily understood when viewed within these categories.

Yes, there are numerous ways to cure the shoulder, *spalla*—whole like a ham, boned and flattened, boned and rolled, boned and smoked. You can slip coppa into a casing, you can season it with hot peppers, but it's still coppa. The jowl, *guanciale*, can be salted plain or it can be rubbed with pepper and bay and juniper, it can be smoked; it can be rolled and dry cured; and it can be cured and pestled into a paste, a preparation called *spuma di gota* (see page 221)—but in the end, it is jowl.

Our intent here is not to complicate the subject of salumi by delving into the countless variations and variations of variations found throughout Italy, but rather to simplify it, so that we begin with a common base, terminology and understanding of this ancient craft, and from this base proceed to deepen our knowledge of salumi's infinite and delicious permutations.

## Hog Breakdown: Italian and American

This is a book about meat, mostly pork, and fat and what it can become when you salt and season it and let it dry slowly. Maybe you'll grind it and stuff it into some kind of casing. Maybe you'll brine a whole muscle and hang it up for months. Maybe you'll cook it when it's only partly dry. Maybe you'll get so excited you won't be able to wait and you'll braise it and then sauté it until it's crispy. Regardless of what you decide, the first and most important thing, as with anything you cook, is to consider the quality of the main ingredient.

As Paul Bertolli, founder and "curemaster" of Fra' Mani Handcrafted Foods in Oakland, California, who makes some of the country's best salumi, puts it, "What makes salumi great is the quality of the meat." That, and how the salumi maker handles that meat: "maintaining cold temperatures; proper butchery, including trimming and tendon removal; use of hard versus soft fats; and then the all-important aspects of cutting/grinding, mixing, and stuffing." In the end, Bertolli says, the complex art of salumi "is a unique transformation wherein seasoning, the by-products of fermentation, mold ripening, drying, aging, and the cellar environment are all 'ingredients' that work synergistically."

And it begins with the quality of the meat. You can't make a silk purse out of a sow's ear, but if you can find a good ear, you can make it fabulously crispy and delicious. It's difficult to make a delicious pork cutlet with one from a commercially raised hog, but cook and eat a cutlet from a hand-raised Berkshire or Duroc hog, and you'll know what pork really is.

We realize that not everyone has easy access to affordable, locally grown Duroc hogs or grass-fed beef and pasture-raised lamb, but some do, and if you do, buy it. If you don't, try to find the best meat in your area. (Or see Sources, page 267, to order over the Internet.) Ask the guys in the meat department at your supermarket where the beef and pork come from. If they don't know, ask the manager of the store. If no one knows where the meat comes from, ask yourself if you really want to shop there. Most metropolitan areas have at least one independent grocery chain, with a local owner. These stores have a little more flexibility in what they're able to bring in. Work with them to find what you want. Then learn how to work with the meat department to get the cuts you want. Educate yourself and become a better shopper, and you will become a better cook.

We emphasize this upfront because the quality of meat is never more important than when it's for a dry-cured preparation. One of the chief pleasures of a slice of dry-cured ham (prosciutto), or beef loin (bresaola), or pork loin (*lonza*) is the intense concentration of flavor it delivers. When meat is dry cured, it loses 30 percent of its weight in water; like a stock reduced by one-third, meat reduced by a third is rich, dense, and intensely flavorful.

As concerns for our well-being, the health of our land, and the facts of commercial meat production raise ethical, health, and humanistic concerns about the animals we rely on for food, more and more people are seeking out meat from humanely raised and antibiotic- and hormone-free animals. This means more farmers are now able to make a living selling such meat. As these farmers grow in number, they're able to sell to more people, increasing demand—a fortunate self-perpetuating cycle.

This is a remarkable circumstance for two Midwesterners who grew up in the 1960s and '70s, when processed and convenience foods were hailed as great advances. If I had asked my dad then if I could buy a hog, my dad would have looked at me as though I'd spoken to him in Etruscan. Same with Brian. The question would have been incomprehensible to a 1970s suburban father.

Today, though, amazingly, happily, I can call or e-mail right this minute any of several farmers who would happily have a whole or half hog for me in a few days. Brian, as a chef, has even more options. A farmer raising Mangalitsa hogs, prized for their copious fat and only recently brought to this country, lives not far north of his restaurant. With the advent of social networks and new possibilities for crowd-sourcing, more and more people should be able to obtain whole animals.

We highly recommend buying whole animals if feasible, and we cannot recommend it more strongly if you're making salumi and other dry-cured preparations.

In fact, we find it difficult to recommend you put in the time and effort to cure your own meat if it comes from a factory hog, which is about all that's available in most grocery stores.

I last bought a hog in Ohio for $1.65 a pound hanging weight (whole but eviscerated): twenty-four chops, two tenderloins, two bellies, two shoulders, two picnic hams, two big back hams, the trotters, and the head. It worked out to about $3 a pound for the meat alone (not including bones, skin, head, and other usable parts). So it's not that expensive . . . *if* you know how to make use of the whole animal. This is important for more reasons than just avoiding waste. It addresses an issue critical to the production and consumption of all our meat. America has become used to the expensive tender cuts of meat, which come from the saddle of the animal, the back from shoulder to above the rump, a relatively narrow section. We are less adept at cooking the rest of the animal, and so there is less demand for these parts (not to mention the offal), and more waste. Learn to cook the whole animal, and you not only become a better cook, you help the animals and help the environment, by feeding more people with fewer animals. It's a big deal.

So, if you want to buy a whole hog to share with several, here's how you might look at it. If you were to split a hog among four families, each family would, for between $50 and $60, take home 3 pounds of belly, to make into bacon; 6 pork chops; a 3-pound pork loin; 5 pounds of shoulder; 3 pounds of picnic ham; and 10 pounds of back ham (to grind into sausage, use for stew, brine for traditional ham, or dry cure, or any or all of the above). You could split the tenderloins, or two families could take the tenderloins and two could take the trotters and jowl meat. Each family would end up with roughly 27 pounds of farm-raised pork at about $2 per pound. And that doesn't include the back fat for sausages, the head for headcheese, the bones for stock, and the skin for cracklings and as a source of gelatin for stocks and stews.

How can you take advantage of the extraordinary bounty that the hog offers us? You need to be able to break it down yourself and know how to make use of it once you do so. If you're buying a whole hog for yourself, you'll also need plenty of freezer space or plenty of space to hang the meat to dry. And you'll need to block out several hours over a few days to do the work, preferably with a couple of friends or family members.

To you cooks at home who want to dry cure your own sausages and whole cuts, if you are going to put in the effort, it's imperative that you seek out quality meat. For those who don't have local access to quality pork, see the Sources list on page 267 for suggestions for ordering pork by mail.

For chefs who want to make dry curing a part of their restaurant's production, the same rules apply, but even more so. Your buying power should give you access to whatever kind of hogs you want. As far as regulations are concerned, as long as you sell your salumi in the place where you make it, you are beholden to the codes of your local health department, not the USDA. Work with the inspectors to educate them about dry curing—how lowering pH, reducing water activity, and using curing salts address any safety issues—so that you can set up a protocol for making and serving salumi that satisfies you and your health department.

Brian got the health department off his back essentially by being more knowledgeable than they were. He addressed exterior bacterial issues with the salt cure. He argued that by using a commercial starter culture, he was reducing the pH to 4.9 or below (the Bactoferm manual is available for download at the Chr. Hansen website), and that he was sufficiently reducing the water activity in the product by drying it. He explained that these techniques have been approved by the USDA for retail dry-curing operations throughout the country. We hope soon there will be a widely accepted protocol.

Again, we cannot overstate the importance of using meat from humanely-raised and slaughtered animals. It should really be the first step of every meat recipe in this book: *First, locate the best-quality meat that you can.*

## Hog Breeds

Specific breeds of hog are used for salumi in Italy. For the famed prosciutto di Parma, Landrace and Large Whites are used because their hams are so big. In Zibello, at the restaurant of the best-known culatello maker, Massimo Spigaroli, we tasted a flight of culatello from Mora Romagnola, Nera Parmigiana, and Cinta Senese hogs.

The last is an heirloom breed we saw in central Italy that nearly became extinct but was brought back in the 1990s by small farmers who now raise them in sustainable numbers. They're adorable, with a single curl to their tail; a wide stripe at their midsection, including the front leg; and ears that flop over their eyes.

In England, Tamworth, Large Black, and Large Whites are the choices for dry curing.

In the United States, Brian and I like Berkshire, Duroc, and Gloucester Old Spot. All of them are easy to raise; are friendly, happy hogs; grow relatively quickly; and have excellent meat, excellent fat, and an excellent meat-to-fat ratio.

A new breed of hog was brought into America recently and is being raised by two farmers we know of. It's an old hog, native to Hungary, called Mangalitsa but also known as the Wooly Pig for its heavy coat of gray fur. It's the fattiest hog we've ever encountered and is used primarily for its back fat, belly, and jowl. The back fat is at least 3 inches/7.5 centimeters thick. The pancetta and guanciale are that thick as well and have very little meat. Indeed, these creatures seem to be composed almost entirely of fat; we're not sure how they even stand up.

Cured Mangalitsa is available from Salumeria Biellese (see Sources, page 268).

Ask your source what the breed of hog you're buying is. Salumi can be made from any breed, and many are now mixed breeds in America. It won't change how you use the hog, but it's useful to know when comparing differing breeds over time in terms of their meat and fat.

## The Experience of Breaking Down a Whole Hog (Is This Your First Time, Sweetheart?)

It's a big deal. Few acts as a cook put you as face to face with the fact that you are eating an animal who lived and died for you and the farmer who raised it as buying and cutting into parts an animal bigger than you are. Sentient creatures. The last one I did had a bright smile and blonde eyelashes. We kill them for food; it's important to acknowledge this by not wasting anything.

It's essential to be prepared. Breaking down a whole (or half) hog is a lot of work, and if you're not prepared or don't have enough people to help you, put your knife down and go home. And if you are not prepared, if you have a feeling bone in your body, you will experience the deep humiliation of having wasted a creature's life because you were lazy. Inexperience is one thing, and inevitable; you can only get experience by starting inexperienced and diving in. Unpreparedness is different—it's avoidable.

Again: a whole hog is bigger than you, if not twice your size, and it's a lot of work. Prepare for it as you would an athletic match. Go through what you imagine will happen in the perfect circumstances you will create for yourself and your fellow butchers/cooks.

Use the illustrations that follow to identify the cuts you want, and have a plan for what you're going to do with them: what you will cook, what you will salt, what you will wrap and deal with tomorrow. For example, we start the headcheese when we begin to work and we cook one of the tenderloins that day for those doing the work. We refrigerate the shoulder meat and trim to dice and salt tomorrow for sausage. It's freezing outside, so we can store all the bones in the trunk of a car and roast them tomorrow. We salt and cure the belly immediately. We portion one loin and freeze it, and we salt cure the other loin. We refrigerate the hams; tomorrow we will divide the muscles of one to brine for cooked ham and salt the other to cure. Whatever you wish. Just don't be standing there staring at all that animal like you didn't expect it to show up on your cutting board.

Drink water. Seriously—you can forget and you'll work better if you're hydrated.

You'll need at least three people who share your work ethic and desire to do this; six is better and more fun. Nine is a party, but save the beer until you begin cooking and all the other parts are taken care of, on the salt (salt begins preserving immediately and is your friend) or wrapped and refrigerated.

Plan to spend several days in all on the work.

A meat saw is like a hacksaw; its teeth are angled so that it cuts only when you are stroking the saw forward, to ensure safety. It should be made of stainless steel, with a plastic handle for easy cleaning.

Tools you'll need:

- Two sharp knives, a chef's knife and a boning knife, and a steel (a sharp knife is a safe knife; hone your knife continually as you work)
- A meat saw (these are really for cutting through bones, not meat, but I guess calling it a "bone saw" in your kitchen would be kind of harsh; see Sources, page 267)
- Three big sheet pans and a roasting pan (or other containers to hold the cuts as you work)
- Big (2.5-gallon/9-liter) zip-top bags for holding and storing the cuts

- Salt: You can salt anything when you finish cutting it. Either a little, as you would season a steak or a roast, if you're cooking it fresh (1% by weight is a good starting point if you need a benchmark; 1% to 1.5% for sausage meat; 3% for all-purpose curing).

One of the most critical tools is your work surface. Do you have a work surface that's at least 4 feet long and 2½ feet wide? That's about the area you'll need at a minimum. A 100-pound half hog is about 5 feet long, so a 6-foot-long surface is optimal. You'll also need space for your tools and trays. If your countertop is not big enough, you'll have to drape a table with an appropriate covering. Most of the cuts you make won't go down to the work surface, but some will, so if you don't have a big wooden butchering surface, you'll need to maneuver a cutting board or two or three beneath each area of the hog as you work on it.

When you get your hog, inspect it. Evaluate it. Does it look as if it's been handled well? Is it clean? If it's not to your liking—if there are gashes in the skin, if it looks excessively bruised—it may have been carelessly slaughtered. You don't have to pay for it. It's not likely that you're going to tell your purveyor no at this stage unless it's a disaster, but just be aware. Always evaluate the quality of your materials.

Use a sharp knife. Find a good wet-grind sharpening service and have your knives sharpened twice a year, or buy a good sharpening device (see Sources, page 268, for the one we recommend). Sharp knives make your work better and easier. They are also safer; you're less likely to cut yourself with a sharp knife (and if you do cut yourself, the cut will be cleaner). Learn how to use a steel to maintain the edge, and use it continually.

Use the right tools for the job: knives cut meat, saws cut bone.

Set up three sheet pans or other receptacles: one for meat, one for trim fat and skin, and one for discards (remove and discard glands, but save silverskin and other connective tissue, such as cartilage or skin, which are rich in collagen, to add to your stockpot).

Things to remember:

- Cut away from your body, not toward it, and down toward the cutting board.
- When seaming, or separating, muscles, use only the tip and first inch/2.5 centimeters or so of the blade. Use your other hand to pull the muscles apart so that you can see where to cut; you can often separate muscles with your fingertips.
- When separating meat from bone, keep the blade as tight against the bone as possible for the best yield. When cutting through muscle, use long, smooth, deliberate cuts, not short jagged ones.

## THE SKELETAL STRUCTURE OF THE HOG

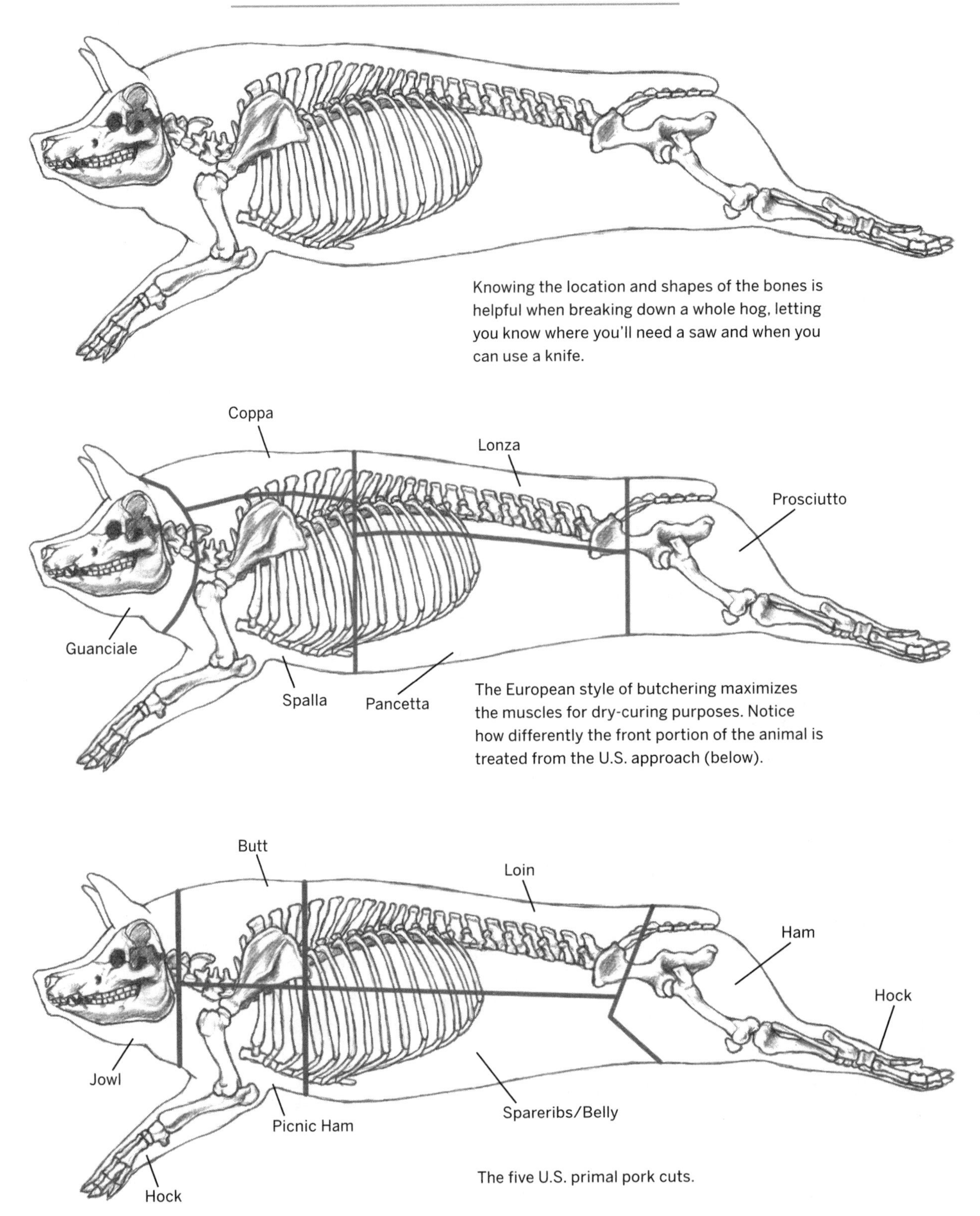


Knowing the location and shapes of the bones is helpful when breaking down a whole hog, letting you know where you'll need a saw and when you can use a knife.

The European style of butchering maximizes the muscles for dry-curing purposes. Notice how differently the front portion of the animal is treated from the U.S. approach (below).

The five U.S. primal pork cuts.

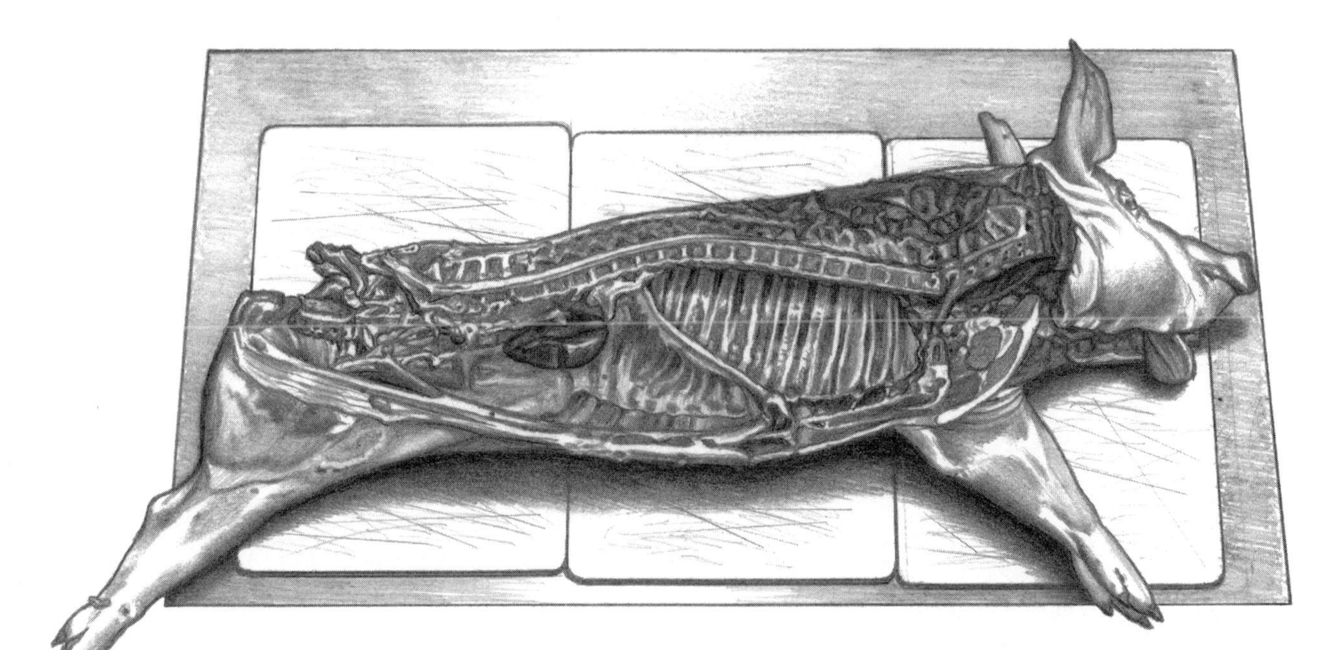

A whole half hog ready for butchery.

## Italian-Style Hog Breakdown for Salumi

There is no one right way to break down a hog. The only wrong way is thoughtlessly or wastefully. Your approach should be guided by logic: how am I going to use the various cuts, how will I cure or cook them? Your answers will determine how you cut the hog in order to maximize the use of this wonderful creature. Every time you butcher, you should have a feeling of pride and respect, knowing that the beast died a noble death for you, and you're happy to be higher on the food chain.

In this section, we describe the Italian method for breaking down a hog for the purpose of making all manner of salumi. It differs considerably from the traditional American method (see page 47) in that it tends to follow the natural seams of the animal with minimal sawing of bone, and it's designed to leave intact the whole muscles prized for curing, such as the *coppa*, the muscle that runs along the neck of the hog.

Much of the reasoning behind the Italian method is practical. For millennia, butchers worked solely with knives. They didn't have electric band saws, and even a good saw would have been a luxury. So the butchers used the natural seams that separate the muscle groups, rather than sawing the hog into rectangles, American-style.

Most of the following butchery can be accomplished with a sharp boning knife, except for splitting the spine to cut the animal in half and separating the ham from the spine and pelvis. It's easier to explain these methods from a half-hog perspective (split lengthwise), which is also the most convenient way to buy the animal (see the illustration on page 33).

STEP ONE

### Head and Jowl

If the kidney is still attached, remove and reserve it; also remove any loose heavy fat. (See kidney preparation and recipe, page 265.)

Remove the jowl by making a cut from the "end of the smile" part of the mouth to just under the eye, continuing back to just under the ear (see the illustration on page 35). Be careful not to cut into the shoulder muscle. Envision a line more or less straight down from the ear, and cut straight down from the spine to the cutting board, separating the jowl from the shoulder. Starting at the ear, peel the fat and muscle back along the skull, cutting slowly and using just the tip of the knife, scraping the cheekbone to leave as little cheek and jowl attached to the skull as possible. Inspect the jowl for and remove any glands, yellowish or grayish nodules that can be as big as a large coin and are squishy like fat but are bitter and undesirable (also keep an eye out for glands in the shoulder area).

If the head is attached, remove it at the atlas joint, where the head connects to the spine. Imagine a line from ear to ear, move your knife 1 inch/2.5 centimeters toward the tail, keeping it parallel to the ears, and make a cut down to the spine. The atlas joint is U-shaped, making it a little difficult to separate the head; take your time, and use the tip of the knife to locate the intersection where the head meets the spine at this joint. Move the knife at different angles until you are able to separate the joint and remove the head.

## REMOVING THE JOWL FOR GUANCIALE

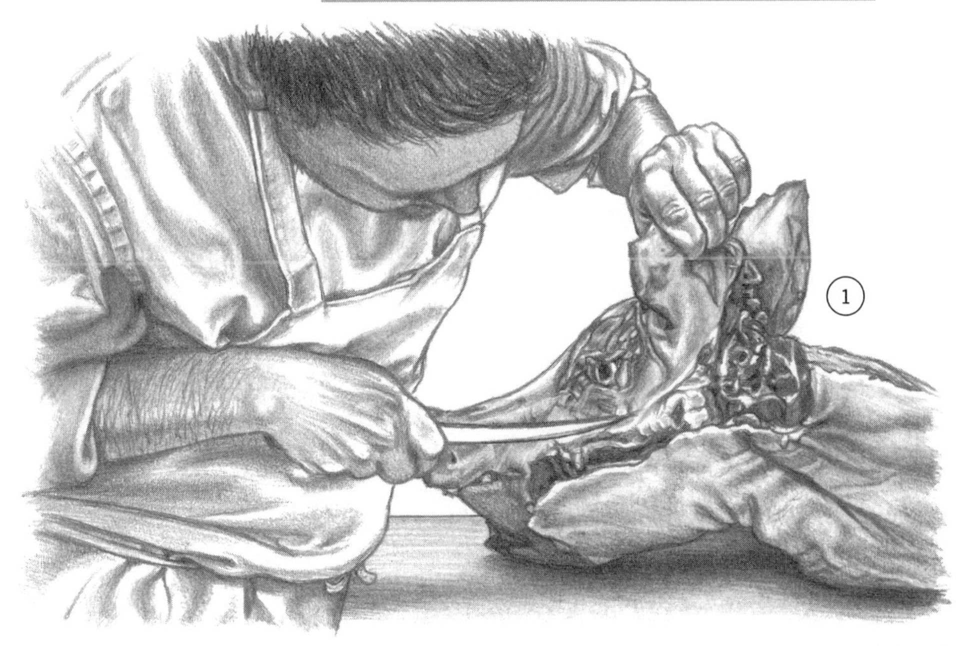


1. To remove the jowl, start below the ear. Follow the bone, making sure to remove as much meat as possible.

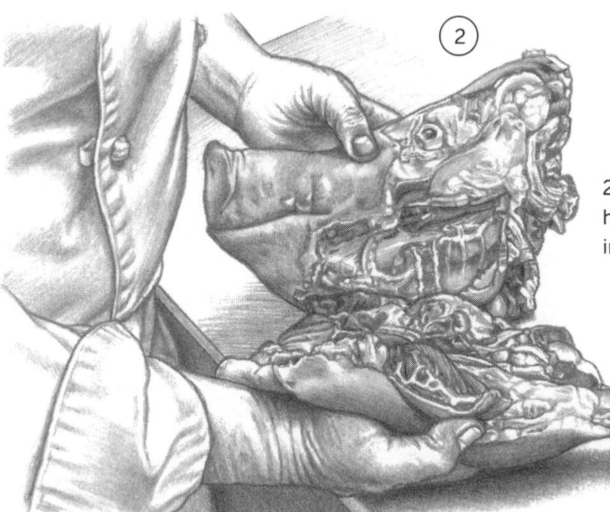


2. The size of the jowl will depend on how the head was removed, but be sure to capture the all-important cheek muscle.

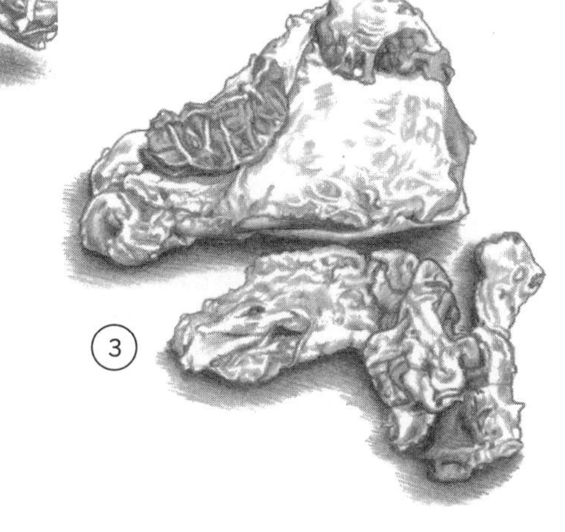


3. The jowl is full of glands; they have an off-yellow or grayish hue and a soft, squishy texture, distinct from the fat. To remove them, trim them away with a sharp knife. Here you can see the glands found in one large jowl.

STEP TWO

### The Shoulder

Turn the animal skin side down so you can see the rib cage. Starting from the front of the spine, count down six ribs and saw through the spine between ribs six and seven (see the illustration below), then continue cutting smoothly through the shoulder and belly. Depending on the age of the animal, you may be able to cut through the breast-plate at the bottom edge of the ribs with the knife. The younger the animal, the more this area is still cartilage; as the animal ages, the cartilage turns to bone, so for a mature hog, you'll have to saw through the breast plate.

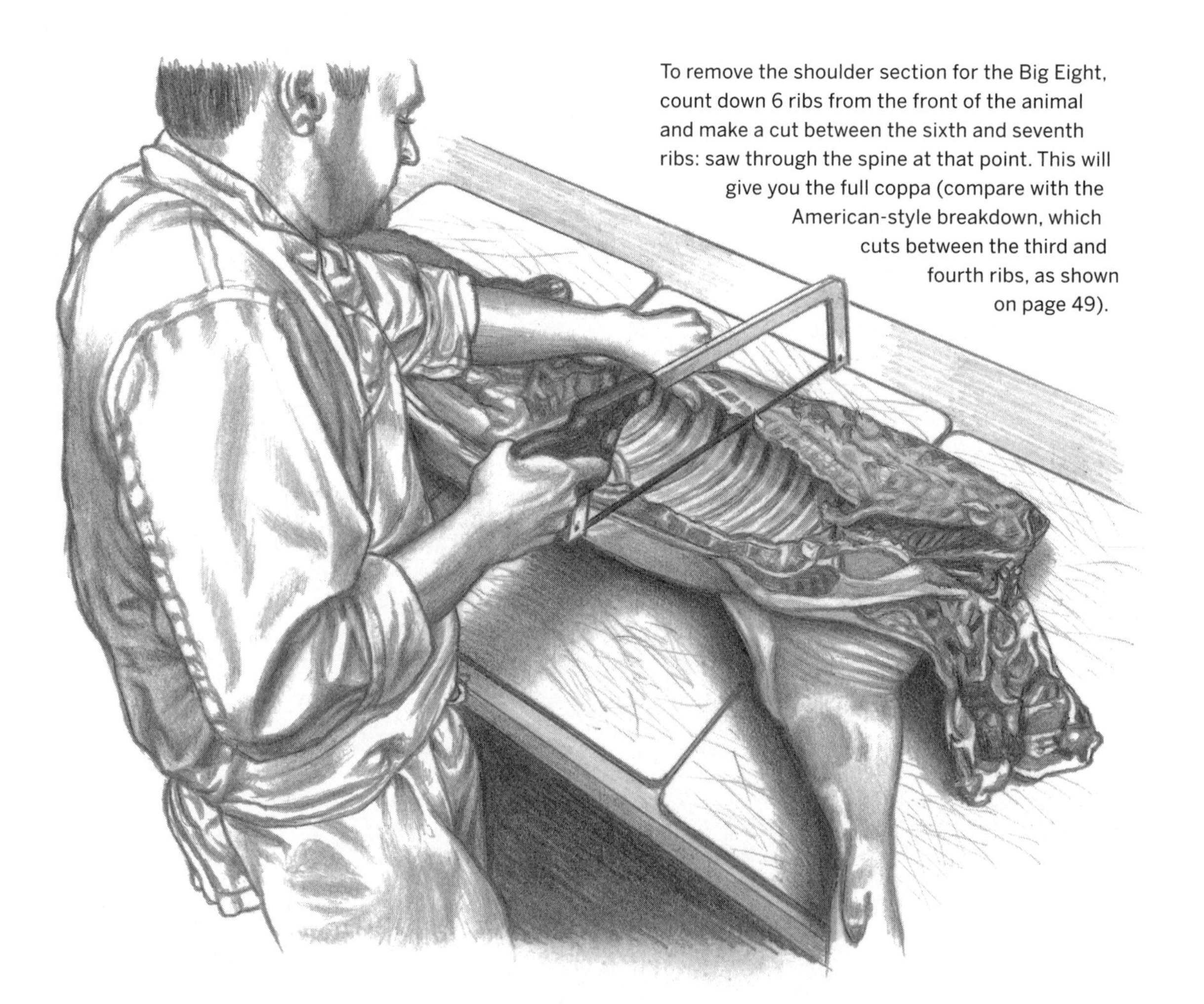

To remove the shoulder section for the Big Eight, count down 6 ribs from the front of the animal and make a cut between the sixth and seventh ribs: saw through the spine at that point. This will give you the full coppa (compare with the American-style breakdown, which cuts between the third and fourth ribs, as shown on page 49).

STEP THREE

## The Ham

Separate the hind leg from the belly and loin by sawing through the aitch bone, the connection between the femur and the spine. Saw through the bone 2 inches/5 centimeters or so below the spine at an angle perpendicular to the leg (see illustration 1 below). When

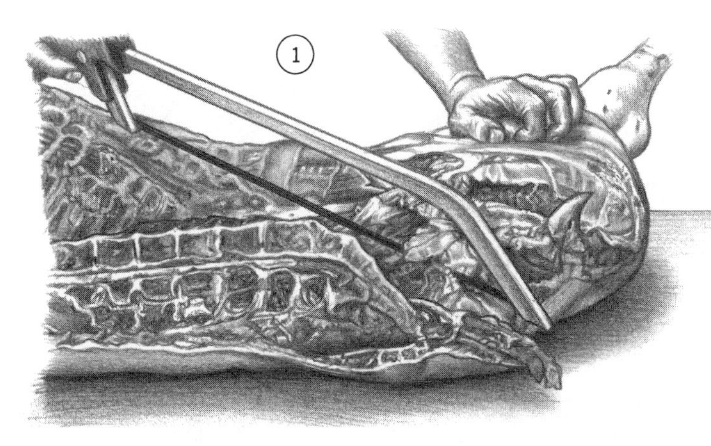


1. To prepare a ham for prosciutto with the aitch bone attached (which we recommend for those inexperienced with removing that bone), saw through the bone halfway from the femur and spine. Remember that saws cut bone, and knives cut meat; as soon as you feel the saw move through the end of the bone, switch to your knife to finish cutting.

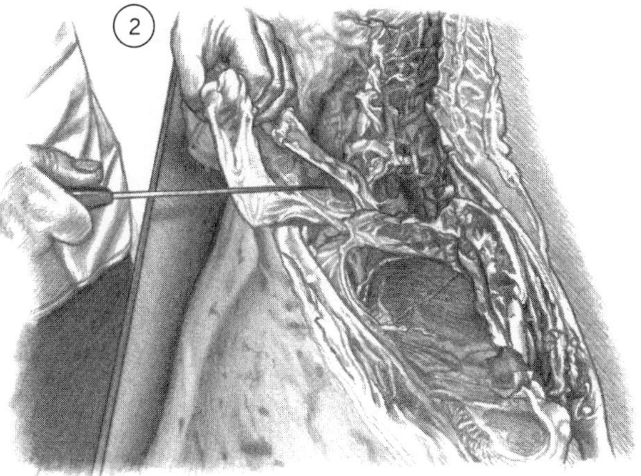


2. To separate the ham from the carcass, follow the natural seam between the belly and the inside round muscle of the ham; you'll find a gland there, marking the separation between ham and belly. Cut through the gland to separate the ham, making sure to remove and discard the gland.

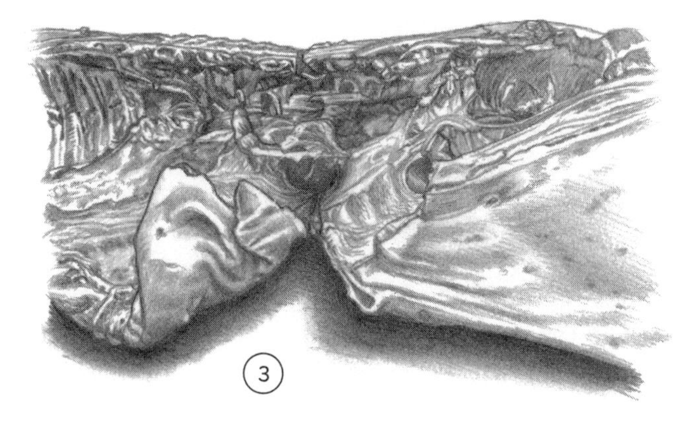


3. To remove the ham from the carcass, follow the natural seam located along the inner muscle (top round), being careful not to cut through the muscle. The seam will guide you where to saw through the spine and help you locate where the aitch bone meets the spine. Saw through the bone, then finish cutting with the knife.

you are through the aitch bone, continue cutting with a knife until you have separated the back ham from the loin and belly, being careful not to cut into the belly. Trim the leg of any excess flap meat and fat (reserve it for salami).

You have now separated the half hog into its three main sections (see the illustration below).

A half carcass broken into the three main sections, ready to be further broken down into the Big Eight salumi cuts.

STEP FOUR

### The Coppa, Pluma, and Spalla

Return to the front section of the hog and remove the ribs and spine in one piece from the shoulder section (see illustrations 1 and 2 on page 39). You can start at the bottom of the ribs or at the spine, slicing with your knife as tight against the bones as possible to remove all the meat there. Be careful: there is a ridge at the base of the spine that can be tricky to navigate. Work your knife around the ridge, capturing all the meat.

## FINDING THE NECK MUSCLE FOR COPPA

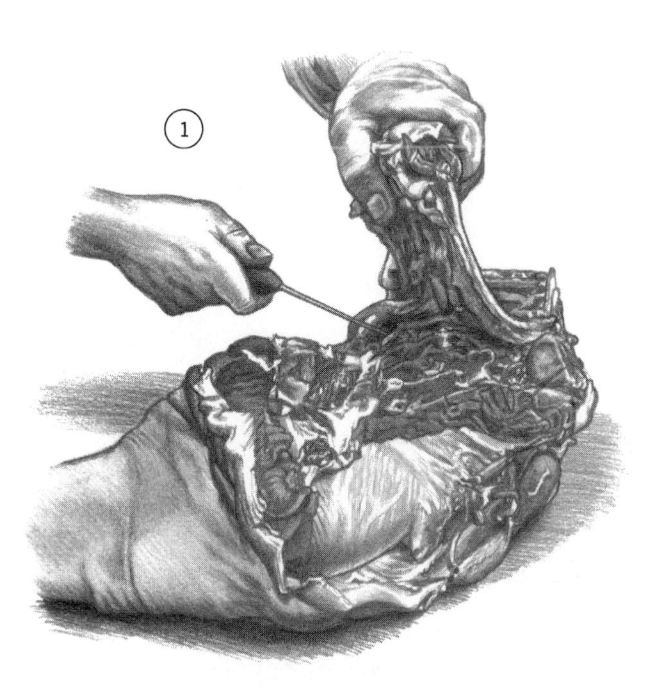


1. The neck muscle (coppa) runs down the top of the spine (it's sometimes referred to as *capocollo*, literally "head-neck"). Remove the bones in one piece to expose the natural seam that separates the coppa from the shoulder.

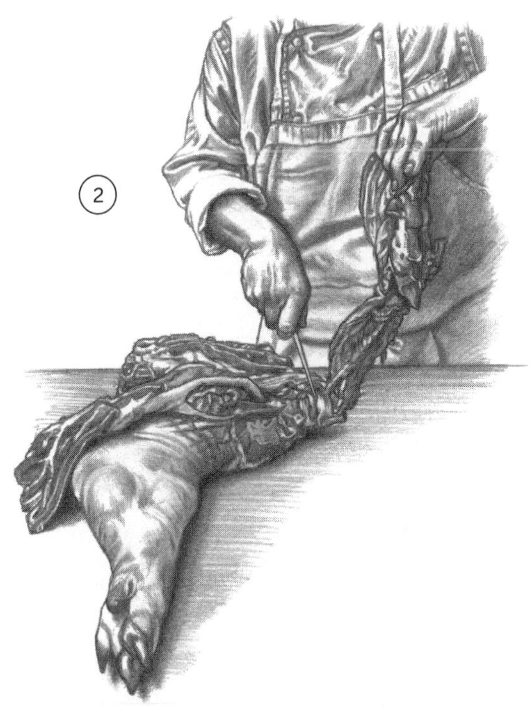


2. Follow the contour of the rib bones and work your way up, releasing the ribs and spine from the shoulder.

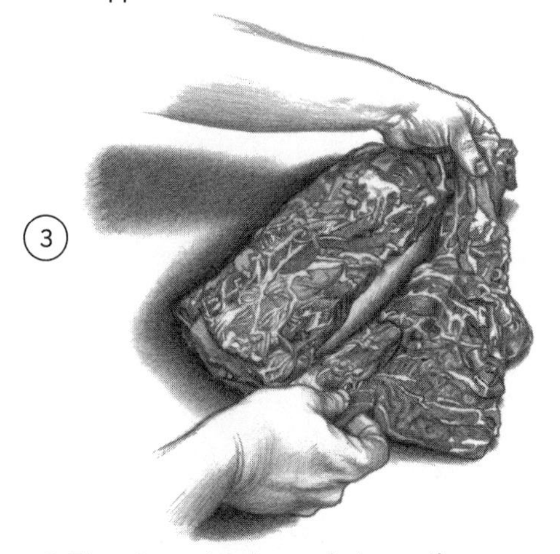


3. There is a natural seam between the coppa and the pluma; follow this seam to separate the 2 pieces of meat. The long loin-shaped muscle is the coppa; the triangular muscle is the pluma (which is typically cooked, not dry cured).

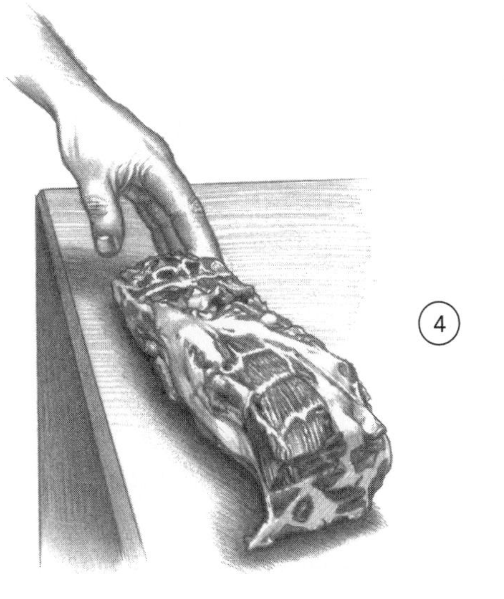


4. The coppa, or neck muscle. Notice the intramuscular fat that makes this such a good choice for dry curing; it is one of the easiest cuts to cure, and one of the best tasting.

Once the rib-and-spine section has been removed, make a cut from the joint of the blade and arm bone, following the seam down to the blade tip and removing the coppa and pluma in one piece. The coppa is the large muscle that looks like beef tenderloin and segues into the loin; the pluma is the triangular flat muscle extending down off the coppa. Separate the pluma from the coppa with a horizontal cut (see illustration 3 on page 39). Trim both muscles and remove the heavy sinews.

The coppa will be cured; the pluma should be cooked and eaten fresh (delicious but somewhat tough, it can be grilled, sautéed, roasted, or braised). The ribs and spine can be reserved for stock or roasted, with plenty of salt and pepper, and eaten as is, as a cook's treat.

What remains is the spalla, the front leg of the hog. It can be cured on the bone or off the bone. If you choose to leave it bone-in, it's a good idea to remove the shoulder blade, which will make slicing the dry-cured meat easier. Remove it by separating the meat from the top of the blade; then, with the tip of your knife, separate the blade from the ball joint of the arm bone and pull that joint end toward you, using your knife to help separate the meat from the underside of the blade (see illustrations 1 and 2 on page 41). Remove the tender "oyster" of muscle in the blade, the *pressa*, which should be eaten fresh, grilled or sautéed.

To bone the shoulder completely, butterfly it to free the meat from the bones (see illustrations 3 and 4 on page 41).

## TWO WAYS TO PREPARE A SHOULDER FOR SPALLA FOR DRY CURING

### BONE-IN

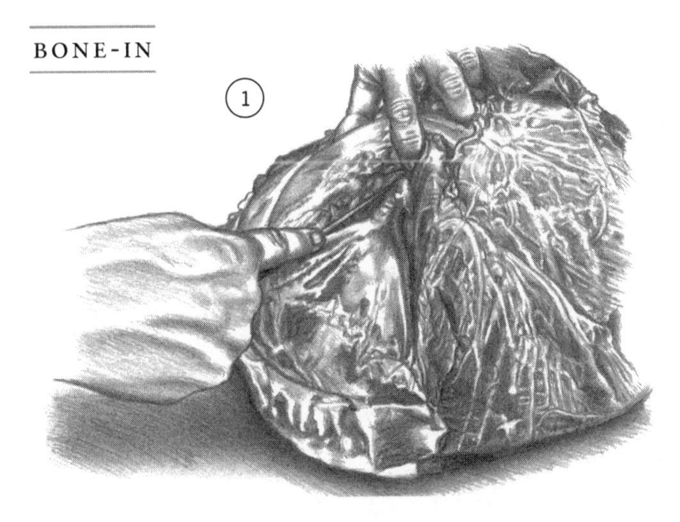


1. To remove the shoulder blade from the spalla: Here Brian's forefinger is shown in the joint where the blade attaches to the arm bone. Find this joint and, with your knife, follow the contour of the bone, separating the muscle from the top of the bone. Follow it all the way around, then cut through the joint.

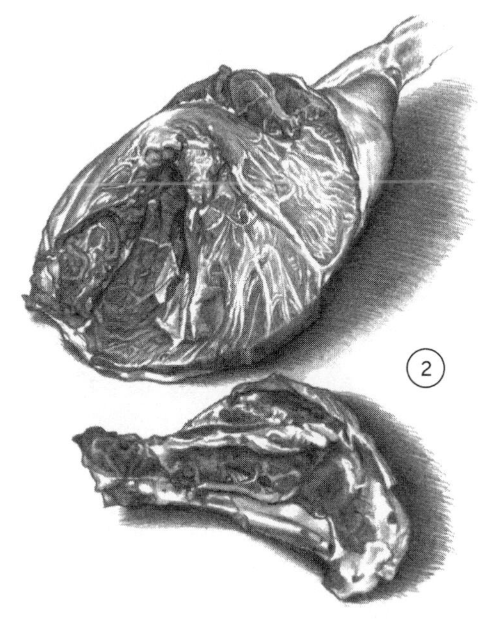


2. The shoulder blade has been removed from the shoulder for a bone-in spalla.

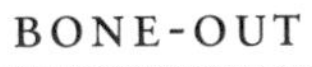

### BONE-OUT

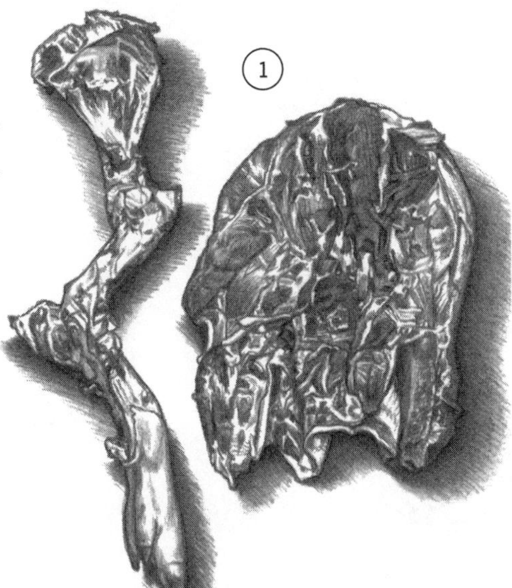


1. Here is an intact front leg (shoulder blade, arm bone, shank, and hoof), beside the boned-out meat. The meat will be used for a boneless spalla (shoulder).

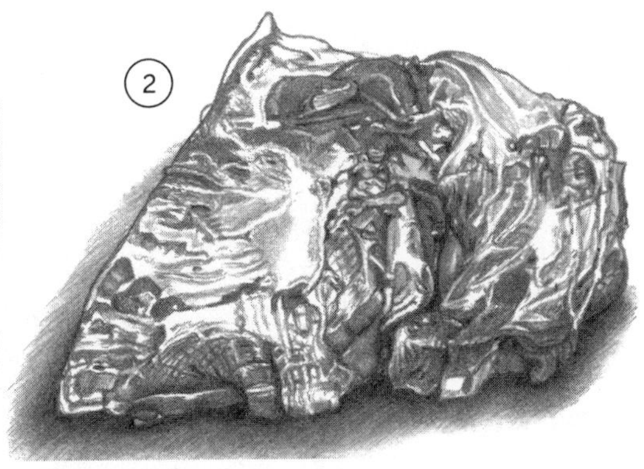


2. A boneless spalla. The arm bone and shoulder blade have been removed, making for easy slicing after it's dry cured. When removing the bones, be careful not to make any deep crevices in the meat, which could encourage the growth of mold.

STEP FIVE

**The Tenderloin, Loin, Belly, and Fatback (filetto, lonza, pancetta, and lardo)**
Remove the tenderloin, which is located just under the back end of the spine (in the illustration below, the tenderloin is just above the kidney; the tenderloin is often the first cut removed—feel free to do this if you wish). Trim the fat and silverskin and set the tenderloin aside; it can be cooked fresh or salted and dried.

Next, remove the rib-spine section: Starting at the end of the rib cage, carefully cut under the rib bones all the way up to the spine, taking off as much meat as possible (see illustration 1 on page 43). When the meat is completely boned, separate the loin from the belly at the natural seam. Along the back side of the loin, there will be a thick layer of fat between the eye muscle and skin. Remove this fat by following the

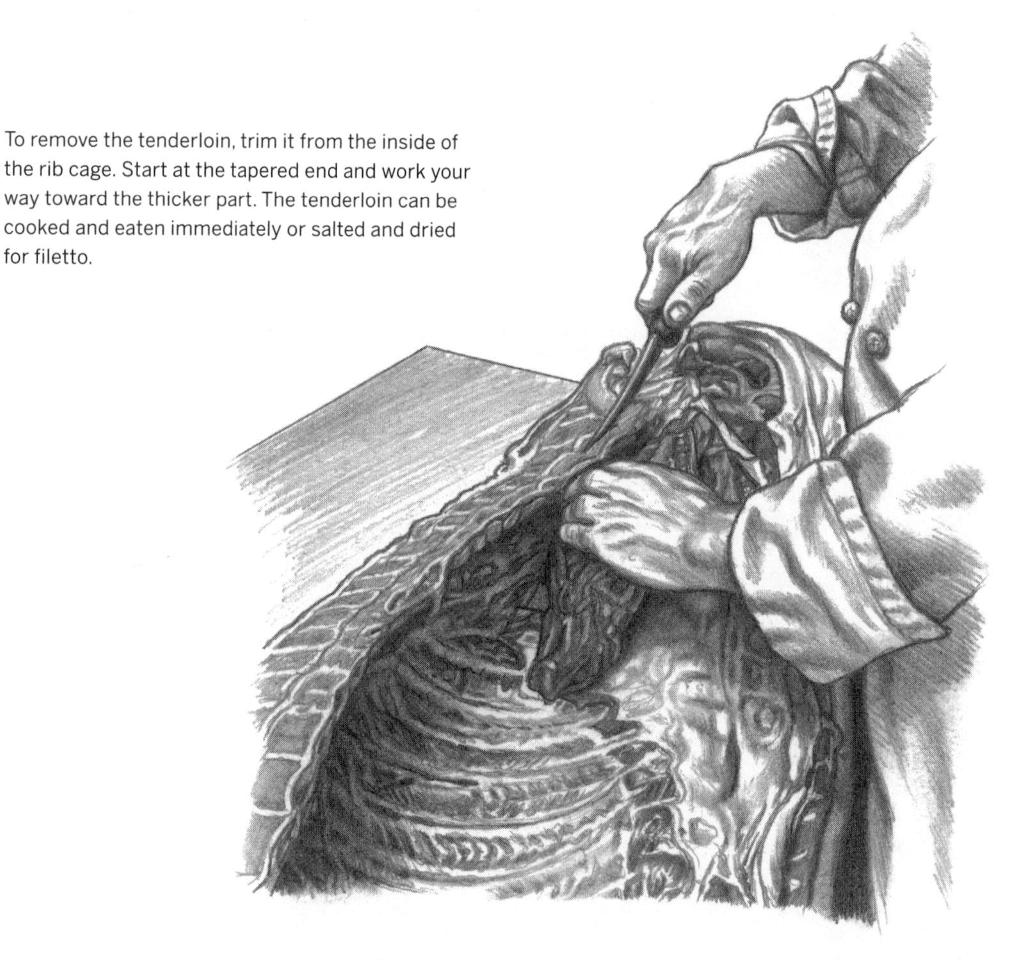

To remove the tenderloin, trim it from the inside of the rib cage. Start at the tapered end and work your way toward the thicker part. The tenderloin can be cooked and eaten immediately or salted and dried for filetto.

## FINDING THE FAT FOR LARDO AND THE LOIN FOR LONZA

1. To remove the spine and rib cage all in one piece, separate this section of the skeleton from the belly and loin; you will see a natural seam that separates the loin from the belly.

2. After you remove the spine-rib midsection and separate the loin from the belly, remove the back fat: with the fat side down, locate the seam that separates the fat from the loin. Depending on the breed, there may be a nice fat layer left on the loin, but in some cases, the seam is next to the silverskin. This illustration shows the famed Mangalitsa breed; notice the thickness of the back fat. Follow the contour of the loin; the seam curves around the loin in a semicircular shape.

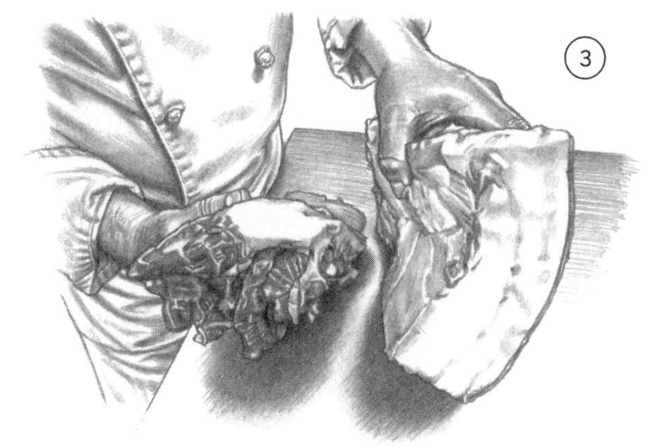

3. Here the back fat has been separated from the loin. Different breeds have different amounts of fat. The Mangalitsa hog is prized for its copious fat and extraordinary lardo.

*continued on next page*

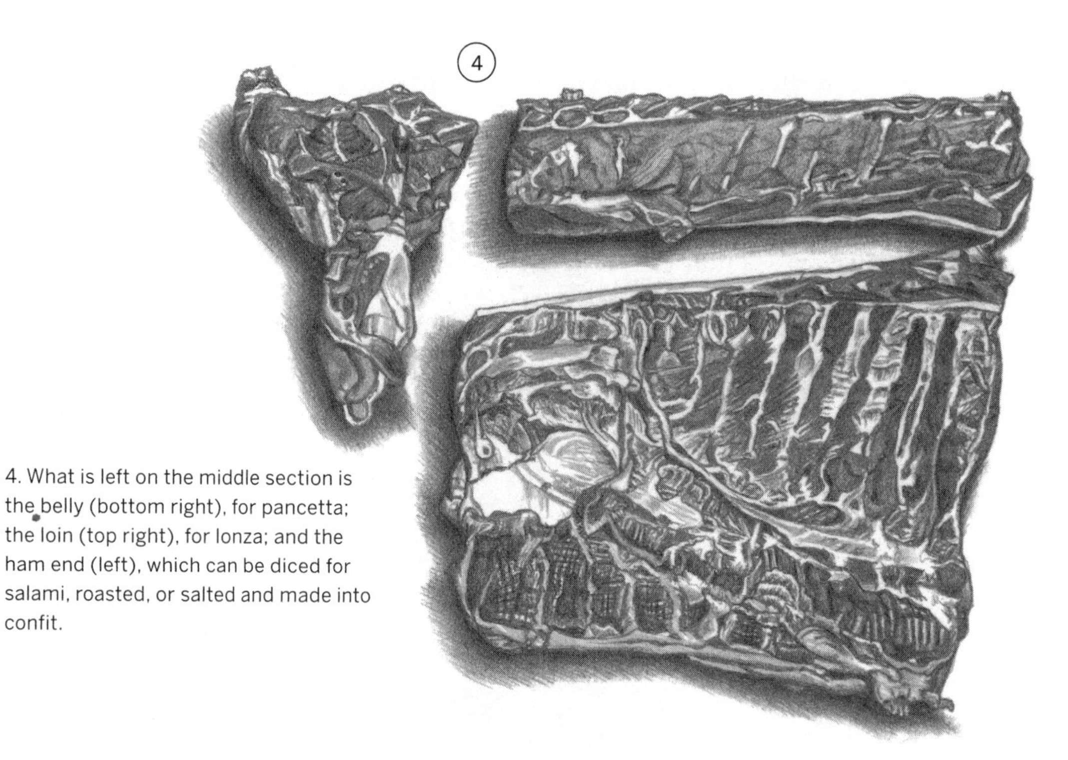


4. What is left on the middle section is the belly (bottom right), for pancetta; the loin (top right), for lonza; and the ham end (left), which can be diced for salami, roasted, or salted and made into confit.

natural seam just above the loin muscle; remove the skin. The loin, belly, and back fat will be cured; the back, a triangular cut of meat, can be cooked or cubed and reserved for salami.

The illustration above shows the loin, the belly, and the triangular back cut.

STEP SIX

**The Ham** (**prosciutto, culatello, and fiocco**)

Traditionally, you would remove the rest of the aitch bone at this point, making the prosciutto easier to slice after it's dried; but for less experienced butchers, we recommend leaving it in to prevent possible air pockets that could harbor harmful mold. If you wish to remove it, cut down along the bone to the ball joint, where it connects to the femur. With the tip of your knife, cut through the tendon between the ball and the socket, and remove the aitch bone (see the illustration on page 45). Trim the prosciutto as desired or as described in the recipe you're using.

If you are making culatello and fiocco, detach the remaining piece of aitch bone at the ball joint and cut it out. Skin the entire back ham down to the trotter (reserve the skin to use in stock for great body or in Tuscan soppressata (page 184). Set the skinned ham inner thigh side up, hoof facing away from you. Starting where the hock meets the shank (about where the Achilles tendon is), make a cut straight through the muscle down to the cutting board and then cut straight along the bone toward you. Work your way around the joint and continue to cut along the bone, the femur, straight through the muscle until the entire back of the ham is separated from the bone (see illustration 3 on page 176). Repeat on the other side of the leg to remove the smaller side of the leg, the fiocco, leaving the bone free of all meat (see illustration 4 on page 176).

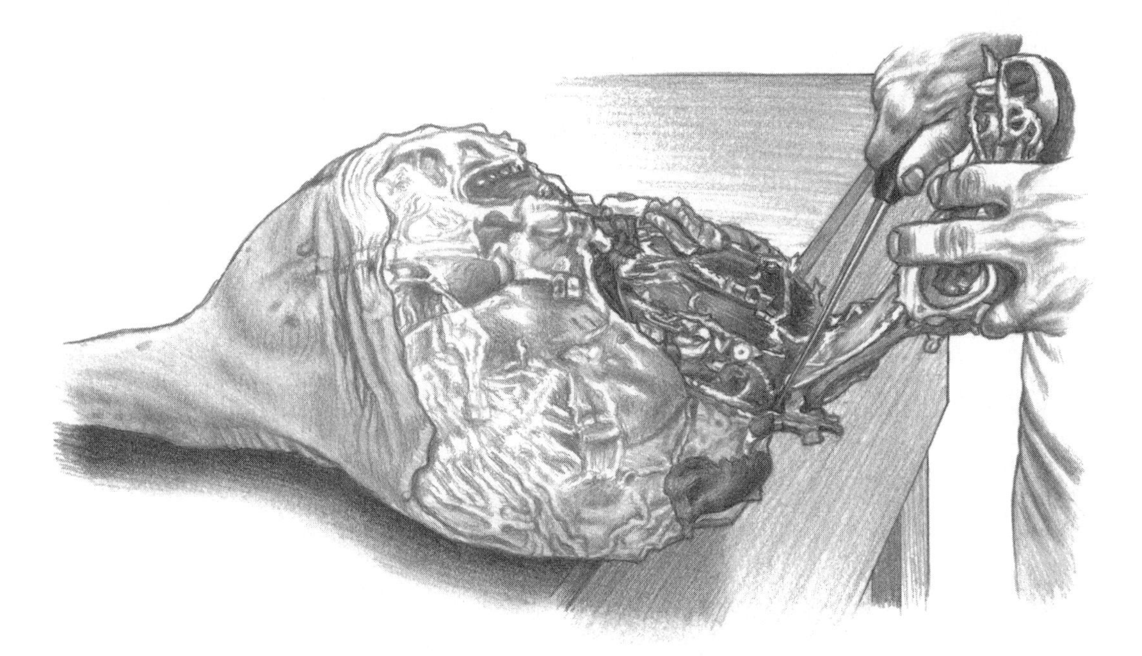

The aitch bone (part of the pelvis, or hip bone) is removed from the ham for traditional prosciuttos. A ham with the aitch bone attached is harder to carve, but removing it improperly can result in spoilage, so, for novices, we recommend leaving the aitch bone attached. If you want to remove the aitch bone, turn the leg skin side down, hoof away from you, and follow the contour of the bone with your knife, making smooth cuts, down to the ball joint. Sever the tendon at the ball joint connecting the femur and aitch bone. Pull up on the ball joint, as shown here, and use your knife to separate the bone from the meat. Trim off any flaps or loose pieces of meat.

## TWO WAYS TO PREPARE THE HAM FOR PROSCIUTTO

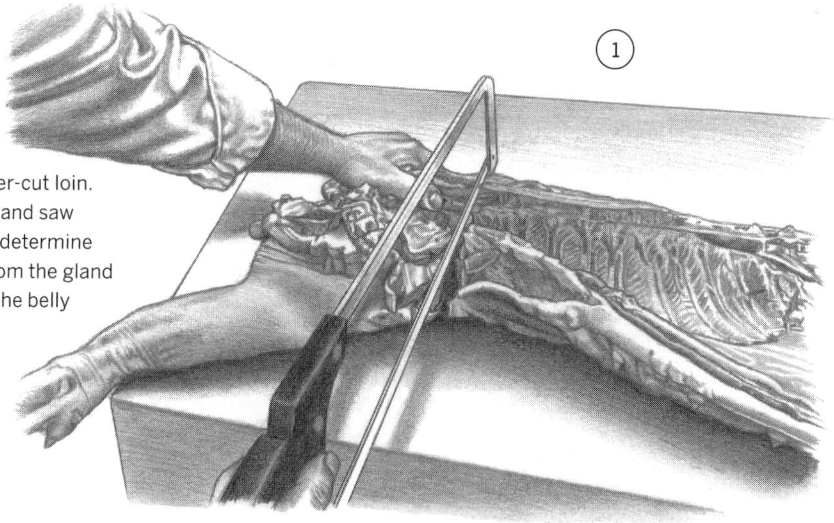

1

1. Separate the ham from the center-cut loin. Count 3 vertebrae up from the tail and saw through the spine. Another way to determine the location is to cut straight up from the gland located at the separation point of the belly and ham.

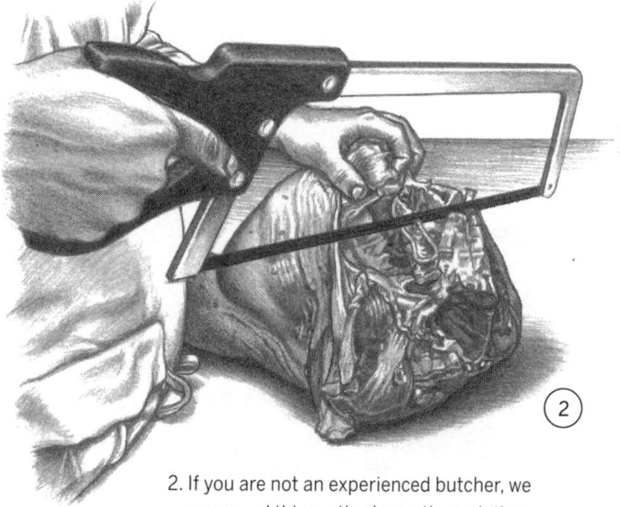

2

2. If you are not an experienced butcher, we recommend this method: saw through the aitch bone rather than remove it. This will square things off and give the ham a neat trimmed surface.

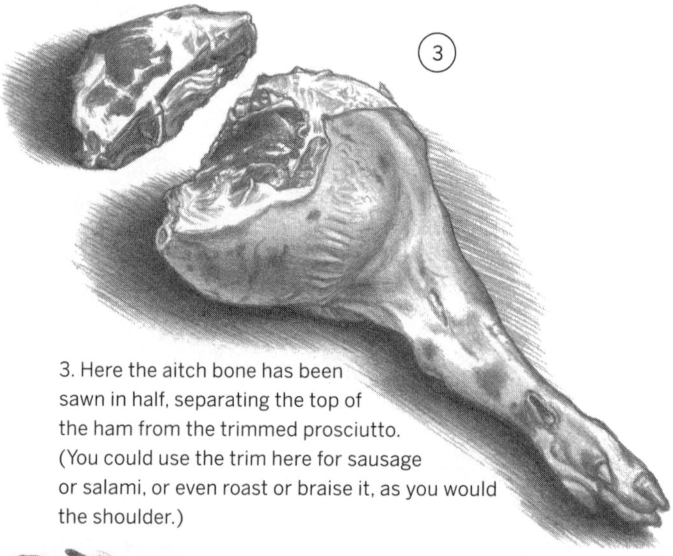

3

3. Here the aitch bone has been sawn in half, separating the top of the ham from the trimmed prosciutto. (You could use the trim here for sausage or salami, or even roast or braise it, as you would the shoulder.)

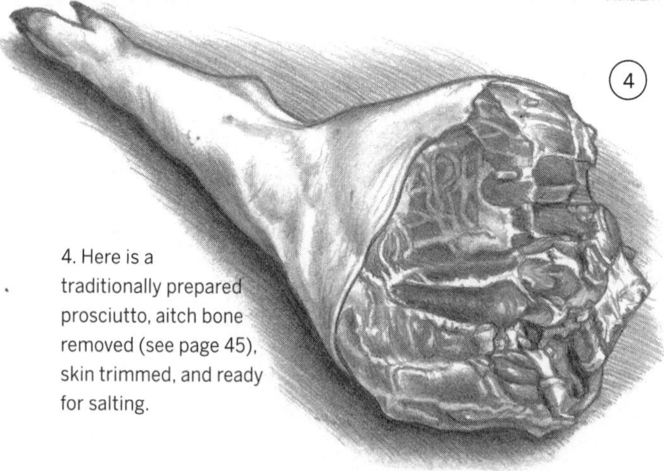

4

4. Here is a traditionally prepared prosciutto, aitch bone removed (see page 45), skin trimmed, and ready for salting.

## American-Style Hog Breakdown

There are many ways of breaking down a hog, each strategy dependent upon how the various cuts will be used. In America, hogs have not traditionally been butchered for salumi. The coppa, the long neck and shoulder muscle that segues into the loin, is not reserved for dry curing; instead, it is cut in half when the head is separated from the shoulder. This type of breakdown takes full advantage of the middle section of the hog for cuts that earn the most per pound: loin, chops, tenderloin, and belly.

We describe how to break down a hog American-style here for two reasons. First, so that the reader who is interested, whether a home cook or a student of butchery and chef, can learn how to do it. Second, because often only the big American primal cuts will be available, and if you want to make salumi, you'll need to know how to extract the appropriate cuts from these parts.

We start by considering a half hog, head removed. (But if you have a choice, ask your purveyor to include the head in your order if you want to work with it.) Most processors can halve the hogs using a band saw, and since sawing through the length of the spine by hand is difficult work, we recommend you order the hog halved and, if possible, with the chine bone removed. After this basic hog breakdown, we address how to handle a whole hog, should you need to.

**The Five Primal Cuts (what you will have when you've finished breaking down a half hog):**

1 picnic ham

1 shoulder butt

1 full loin with back ribs

1 belly with spareribs

1 ham

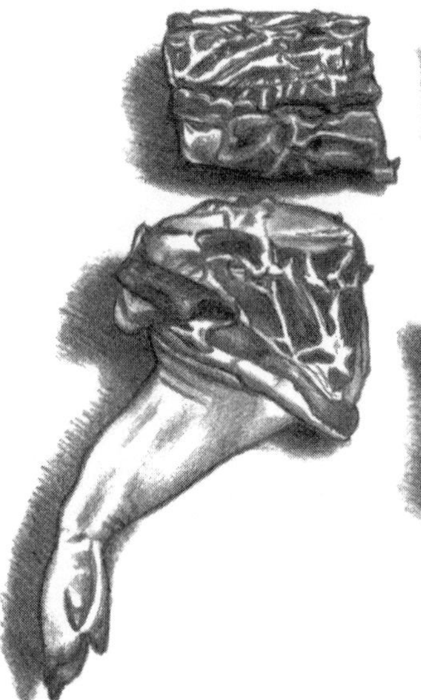

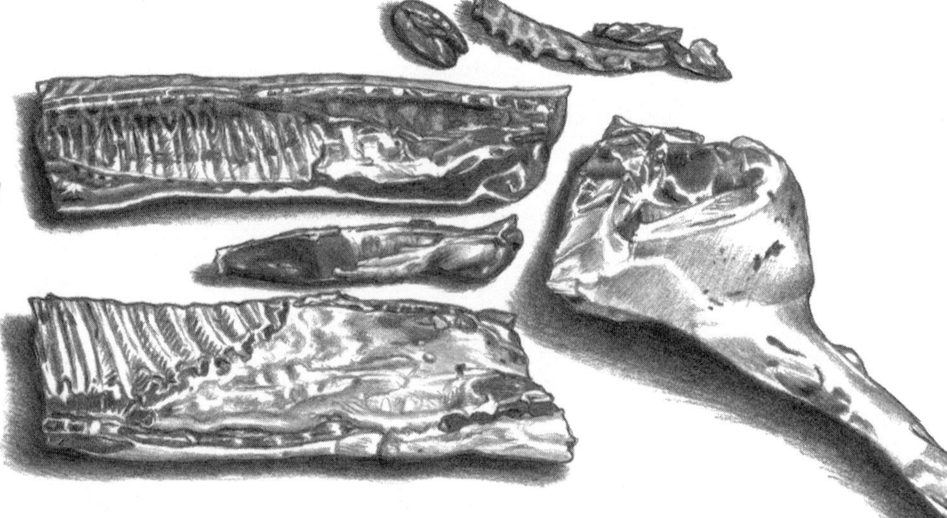

The five U.S. primal pork cuts. From the left: shoulder butt, picnic ham, center-cut loin, belly, and ham. Notice the tenderloin between the belly and loin, and the kidney and hanger steak above the ham.

STEP ONE

### Separate the shoulder and front ham from the rest of the body

With the hog skin side down, trotter pointing toward you, draw a line between either the second and third rib or the third and fourth rib, depending on the size of the animal: the idea is to get as much of the shoulder meat in the shoulder cut without cutting into the loin, the muscle that runs the length of the spine along the upper back of the hog. Another way to determine this cut is to draw an imaginary line straight up from the hog's elbow.

Cut straight down this line, perpendicular to the spine and between the ribs. You'll need to saw bone in two places: saw through the spine and saw through the blade bone, leaving the majority of this bone in the shoulder and about an inch or two (about 2.5 to 5 centimeters) in the loin section. Remember: cut bone with saw, meat with knife. The teeth on the blade of a meat saw angle forward, meaning you are only cutting when pushing the saw forward. As soon as you are through bone, stop with the saw and resume with the knife.

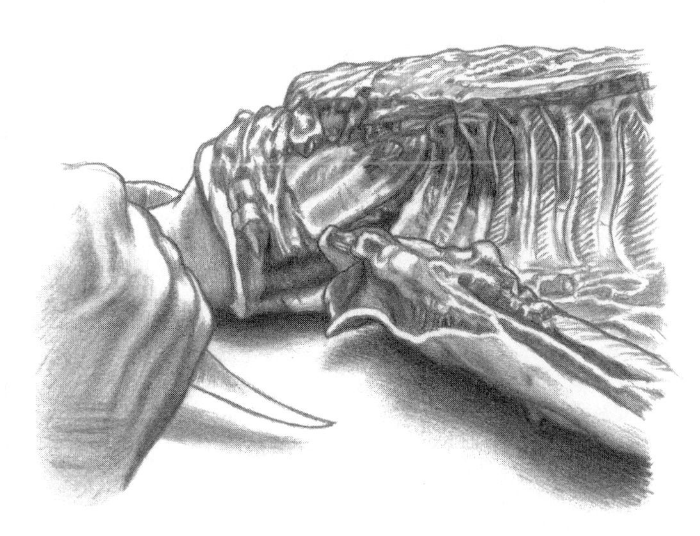

To separate the shoulder from the rest of the carcass for the American-style breakdown, count 3 ribs from the front of the animal and, with your knife, make a cut between the third and fourth ribs. Depending on the age of the animal, the breast plate may be soft enough to cut through with a knife, as shown here. If it's an older animal, you will need to use a saw.

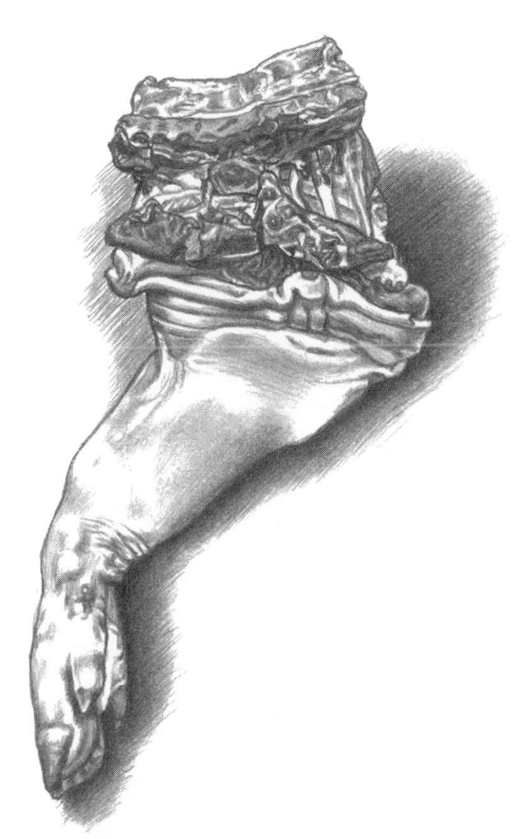

In the American-style breakdown, the front third of the animal is separated from the rest of it by sawing through the spine between the third and fourth ribs. This will be halved, to create the shoulder butt on top and the picnic ham below.

STEP TWO

### Separate the shoulder butt from the picnic ham

Divide the butt from the ham at the midline, gauging it by eye and cutting through the midline between shoulder and ham down to the bone (depending where you cut, you will hit either the tip of the blade bone or the arm bone). First saw through the ribs remaining in this cut. Then cut through the meat until you hit bone. Saw through the bone and then continue cutting through the meat with the knife. This process is called squaring off the butt, because you are cutting at right angles and will end up with a block-shaped portion of shoulder butt. If the picnic ham has the shorter end of the blade in it, remove this piece of bone where it connects with the arm bone. Remove the spine section from the shoulder.

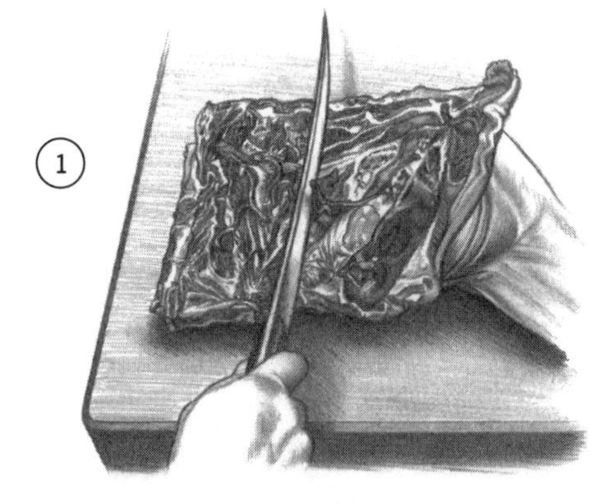


1. To separate the shoulder butt from the picnic ham: with the shoulder skin side down, measure halfway up the arm bone and saw through the arm bone, then finish the separation with a knife. This will yield a squared-off butt and a picnic ham with shank.

2. Lay the muscle fat/skin side down. Locate the shoulder blade before you start cutting. Be sure to make smooth, deliberate cuts so you don't make any crevices in the meat, which could encourage mold.

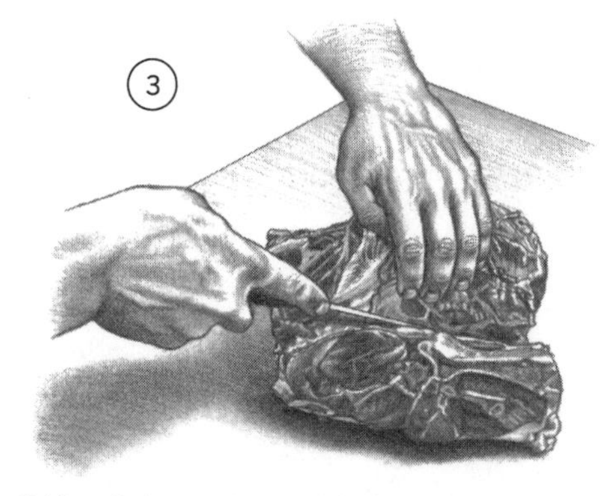


3. Visually locate the coppa muscle in this U.S. primal shoulder butt by standing it on end, you should be able to see the blade bone; one side is flat, the other has a ridge. Separate the coppa from the bone by cutting along the flat side of the bone. With the tip of your knife, find the flat side of the shoulder blade and follow the contour of the bone, separating the cut into 2 pieces.

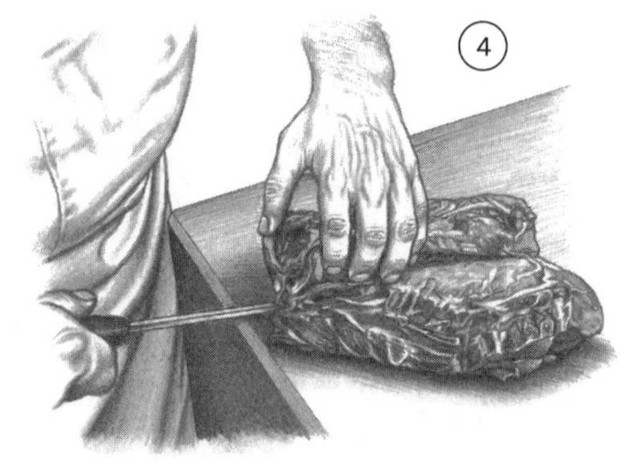


4. Once the shoulder butt has been separated into 2 pieces, lay the boneless piece on a work surface. Find the cylindrical muscle, locate the seam, and separate the coppa from the pluma.

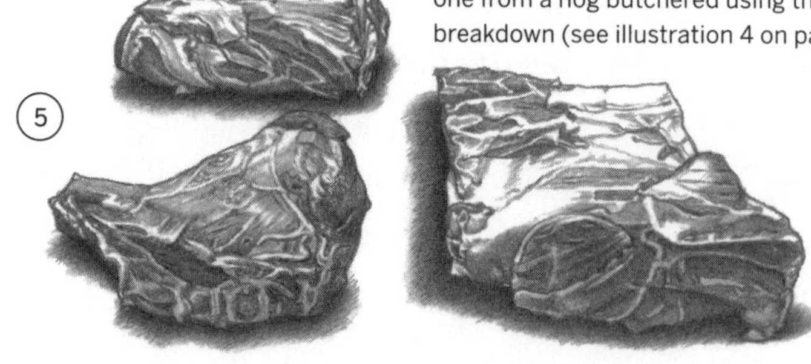


5. For a U.S. pork shoulder butt, the coppa (top left) is removed and separated from the pluma (bottom left). Notice how much smaller the coppa is than one from a hog butchered using the Italian-style breakdown (see illustration 4 on page 39).

STEP THREE

**Remove the trotter from the picnic ham**

Feel for the joint where the hock connects to the foot of the hog and saw through the foreshank bone that connects to the foot bones. You can also separate the trotter from the shank at that joint.

STEP FOUR

**Remove the ham**

Stand at the tail end of the animal. The tail will have three or four joints: Count three joints in and begin your cut, a straight line perpendicular to the leg bone down to where the belly begins. You will be able to see where the belly begins; don't cut into it. Smack in the middle of the line where you will cut is the aitch bone: Feel it with your knife. Saw through this bone first. Make the remaining cuts with your knife, from the tail end down toward the belly; when you reach the belly, angle away from the belly straight down. The ham is now separated from the rest of the body.

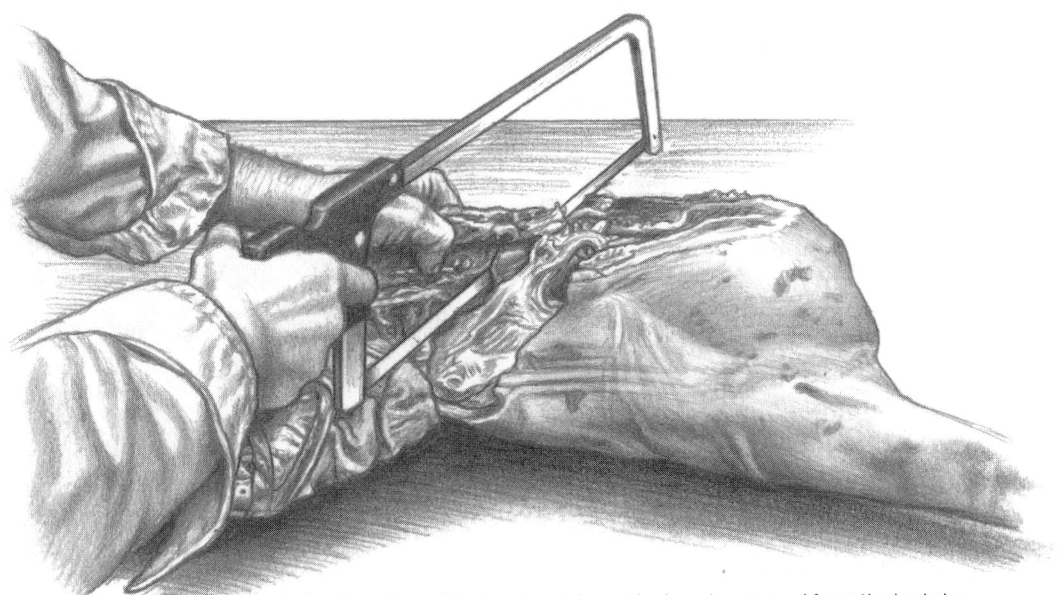

In the American-style hog breakdown, the ham is removed from the body by sawing through the spine 3 vertebrae up from the bottom of the spine (not the tail, which extends beyond the end of the spine).

STEP FIVE

### Separate the belly from the loin

The belly rises about two-thirds of the way up the sides of the hog, tapering into the fat that rises over the loin as the ribs bend in toward the spine. At the front of the hog, identify where the ribs bend and make a mark with your knife about an inch/2.5 centimeters below the bend. At the back end of the hog, you will see the loin and tenderloin. Make a mark just below the tenderloin. Now it's a matter of connecting these two marks with a straight line; this is the course of your knife. Standing at the front of the belly, begin cutting from the rear toward you along this line; about halfway through this cut, you'll reach the ribs. Using a saw, cut through each rib. (If, when you removed the shoulder, you cut between the third and fourth rib, you'll be cutting through twelve

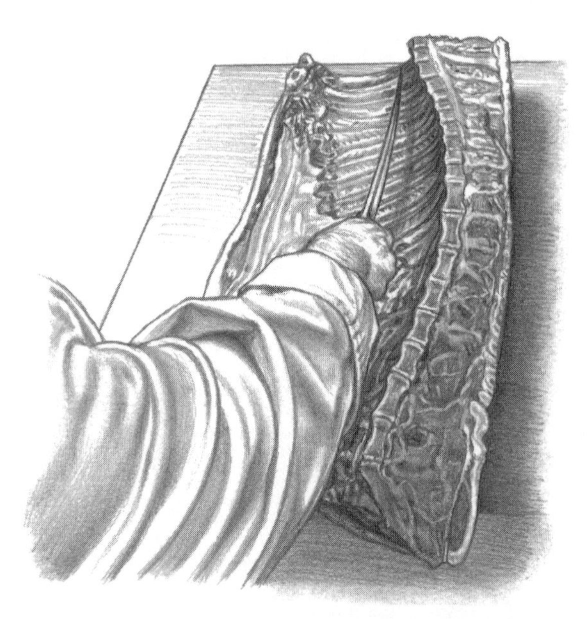

The belly is separated from the loin by scoring a line from one end of the belly to the other, about 3 inches/7.5 centimeters down from the eye of the loin on the shoulder end to 1 inch/2.5 centimeters from the eye on the ham end. Using the saw, carefully cut through each rib, then finish the cuts with the knife. This will leave you with the belly (spareribs attached) and a bone-in center-cut pork loin.

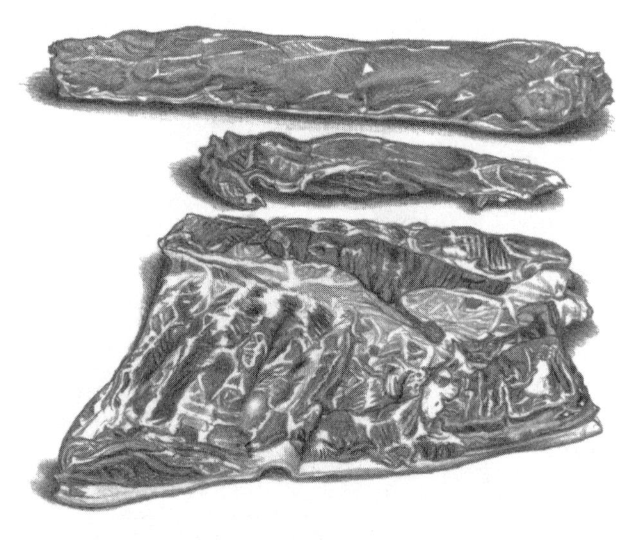

The cleaned meat from the center cuts (from top to bottom): loin, ready for lonza; tenderloin, ready for filetto; and belly, ready for pancetta.

ribs.) Be careful here; you only want to saw through the rib bones, not tear into the meat with the jagged edge of the saw. You have most control over the front of the saw blade, so use that in short strokes to get through each rib. When you have sawn through all the ribs, cut with the knife to separate the large rectangular belly from the loin section.

The ribs that remain attached to the belly are the spareribs. The flap of muscle extending off the ribs is the skirt steak (the diaphragm). The ribs and skirt steak can be removed from the belly by sliding your knife under the ribs to separate belly from ribs. The skirt can be removed and used on its own (awesome for carnitas) or left on and cooked with the ribs. However you use it, remember that it's a tough cut of meat (as are the ribs), requiring long, slow cooking.

### Now What?

So you're standing over your wonderful primal cuts, proud of your skillful knife work. Now, what to do with these five pieces of meat? You have many options. For example, if you're a restaurant chef selling pork chops and other tender cuts, you'll want to remove the tenderloin and cut chops. If you want a whole cured loin or a whole slab of smoked Canadian bacon, remove the loin completely from the bone and trim the fat around it. If you are at home, you'll likely want a little of each, so you might cure the back end of the loin in some fashion and cut the back ribs (on young hogs, these are baby back ribs) into chops.

### The Subprimal Cuts

Here's a general list of the cuts, called subprimals, available to you when you break down your own hog.

THE SHOULDER BUTT can be:

- Ground for fresh or dry-cured sausage
- Cut into steaks and cured like bacon (tasso ham)
- Cured, smoked, or baked, as you would a regular ham

Cured and poached in fat for confit

Slow-roasted for a shoulder roast

Covered and slow-roasted until shreddable for pulled pork, carnitas, or tacos

Smoke-roasted and sauced for pulled pork (barbecue)

Cut into cubes and used in stews

THE PICNIC HAM can be:

Cured and baked

Cured, smoked, and baked

Cured and dried for spalla

Cured, smoked, and dried for speck

Cut up for stew

Cut up for sausage

The trotters can be braised

The hocks (remove them using knife and saw) can be cured, and smoked or braised to add to trotter preparations (see Zampone, page 190)

THE LOIN SECTION: There are numerous ways to break a whole loin section down into parts. First, the tenderloin should be removed and served for dinner: it's a lovely, flavorful muscle that can be pan-roasted whole, butterflied and grilled, or cut into medallions and sautéed, grilled, or roasted to medium-rare to medium. Or, of course, it can be dry cured.

The back fat is wonderful for use in pâtés and in sausages, but it can also be cured as lardo (see page 96). The skin is rich in collagen, which will break down into gelatin, giving stocks, soups, and stews body. When the skin is cooked until tender in a stock or stew, you can scrape the excess fat from it and roast it or deep-fry until it's crisp and eat it as cracklings.

And then there's the long loin section. Half of this rests on the back ribs. You can leave this on the bone and cut it into chops, or you can roast the whole rack of ribs, a festive celebratory preparation. The rest of the loin can be roasted whole or cut into boneless chops.

If you're cutting this section into bone-in chops, either before cooking it or after, you'll need to remove the chine bone, where the ribs connect to the spine. You can't cut through this with a knife. If you know ahead of time you want chops, you might ask the purveyor to cut the chine bone out (much easier with a band saw). But you can do it yourself with a meat saw.

THE BELLY: Remove the spareribs and skirt steak to braise or barbecue separately (see page 53). The belly can then be:

Cured and hot-smoked for bacon
Cured and air-dried for pancetta
Cured and roasted for unsmoked bacon
Cured and confited, which both tenderizes and preserves this cut, for any number of other preparations.
Roasted low and slow until tender
Braised and cooled in the braising liquid, then cut and sautéed, roasted, grilled, or fried

THE HAM can be:

Cured and dried in numerous ways
Roasted
Brined and roasted
Broken down into various muscles that can be used on their own (e.g., dry cured for culatello, brined into small hams, or simply roasted)

**The Other Cuts (should you be so lucky)**

If you have the head, you have a gift. You can prepare a traditional headcheese (see the Tuscan soppressata on page 184), simply by poaching the whole thing until it's all tender and falling off the bone, or you can go four-star, boning it out, rolling the tongue and ears up in the meat and fat and skin, and braising it for a truly spectacular presentation.

If the head came attached, remove it (see page 34) before you begin the other cuts. If you're looking to cure the jowl for guanciale, leave as much of it on the head as possible; the jowl elides into the neck, with much glandy fat separating shoulder meat from jowl (see page 34). The puck of muscle attached to the cheekbone is the hog's cheek, a succulent cut to braise. When you've removed the jowl, trim away the copious fat that contains pale, meat-like glandular tissue. Also remove and discard any pale, squishy, not-quite-meat, not-quite-fat disks—these are glands and no fun to eat. Examine your shoulder meat for these as well, discarding any you find.

## Is It Financially Practical to Buy a Whole Hog?

We recently bought a hog from a local farmer (eviscerated, de-haired, halved, skin and head still attached). About 250 pounds at slaughter, once dressed, the hog came in at 206 pounds. This is a fairly small hog; for copious fat, we recommend asking your farmer to grow it out to 300 to 350 pounds. Our price was $1.90 per pound hanging weight, or roughly $400 for the whole hog (chefs buying in greater volume may be able to negotiate an even better price).

After the complete breakdown, this is what our hog yielded:

2 jowls, 5 pounds/2250 grams each

2 shoulders, 12 pounds/5500 grams each

2 coppas, 6 pounds/2750 grams each

2 loins, 6 pounds/2750 grams each

2 pancettas, 12 pounds/5500 grams each

2 prosciuttos, 24 pounds/1100 grams each

Trim (for salami), 18 pounds/8250 grams

2 tenderloins, 2 pounds/875 grams each

---

Total trimmed weight: 152 pounds/69510 grams

Curing and drying the meat will result in the loss of approximately 30 percent, yielding about 106 pounds/48660 grams cured meat. So the total cost, at $400 for 106 pounds = $3.77 per pound for homemade salumi. (Costs will end up being slightly higher because of casings and seasonings, and the weight of the bone in the prosciutto.) We've gotten it to be as low as about $3 per pound/450 grams. For high-quality hand-raised meat, that's a great price.

From a restaurant perspective, these numbers do not even include the real cost of bones, head, and trim, which are used for stock, headcheese, and other preparations that can be sold.

At Brian's restaurant, he figures about 2.6 orders per pound, and he charges about $8 per order.

Final verdict: 106 pounds × 2.6 orders = 275 portions × $8 per portion menu price = $2204 for a $400 raw-cost investment, or roughly an 18% food cost.

Bear in mind, of course, the space needed; the variable conditions; the quality of the hog, and the labor, time, and talent needed to achieve a high-quality finished product; and the market. But if you are able to account for all of these, using whole animals can be better for the animal, the farmer, and the consumer; can yield more flavorful results; and can also be highly profitable.

# 2. Dry Curing: The Basics



Dry curing means reducing the amount of water in meat to create an environment inhospitable to the bacteria that cause spoilage. Our ancestors learned how to do it so that they didn't waste food, and so that they could create a supply of food. Preserving food was essential for survival. Before the advent of refrigeration and freezing, families that raised whole animals couldn't eat the meat from an entire animal before it began to go bad, so they learned to salt and dry it.

We no longer need to cure meat for our survival. Now we cure meat because it's so delicious to eat. And this is precisely why we find curing so alluring: what was done to the meat originally to keep us alive results in some of the greatest culinary pleasures available to us, a reflection of ingenuity, craftsmanship, and love.

Anyone can dry cure meat if they have the right environment, and some preparations, such as pancetta, don't even require a special place for dry curing (Michael hangs his from a hook in his kitchen).

First, we must note that dry curing *ground* meat in America in the twenty-first century requires the use of sodium nitrate, primarily to protect it from botulism contamination. But sodium nitrate is not essential from a safety standpoint for whole muscles. Many of the best producers in Italy use only salt and seasonings. In this chapter, we discuss the use of curing salts, but in Part One of the recipe section, we describe those pure methods as practiced in Italy.

## Salts

Salt, sodium chloride, is the most important ingredient in the kitchen, but it keeps a low profile by nature. In the salumeria, salt is the magic key, the salumière's primary tool. Without salt, salumi—indeed, virtually all forms of preserved foods—would not be possible. Salt creates an inhospitable environment for bacteria. It also pulls water out of food, which further inhibits the growth of harmful microbes. It helps lower pH levels, further protecting the meat from bacteria. Because it can bind water to the meat proteins, and thus inhibit drying, salt is typically added just before grinding meat for salame.

Because salt is essential to our survival, our bodies are well attuned to its presence in foods—in natural foods, that is. Salt hides out in processed foods, which may be the primary reason salt consumption has become a problem in America. But the problem is that Americans eat too much processed food, not that we put too much salt on our steak or tomatoes. Avoid processed foods, and if you don't have preexisting hypertension or other salt-related conditions, you should be able to salt your food to whatever level pleases you without worry.

To avoid oversalting foods such as dry-cured sausages, we can't recommend strongly enough that you weigh your salt rather than measure it by volume. Different brands have different weights by volume. A cup of Morton's kosher salt weighs about 8 ounces, but a cup of Diamond Crystal weighs between 5 and 6 ounces—more than a 25 percent difference. A cup of kosher salt weighs less than a cup of fine sea salt.

For all dry curing, we recommend Trapani sea salt, an Italian sea salt from Sicily used all over Italy and by many salumi makers in America (see Sources, page 268). It contains trace elements—magnesium, potassium, and calcium—that have some effect on the final flavor. We use it because salumi makers we respect use it, and because it works. It's worth using if you're doing a lot of dry curing, but other good-quality sea salts are fine to use, as are Diamond Crystal and Morton's kosher salts. If you do not own a scale, use Morton's kosher salt; it has an almost even volume-to-weight ratio. (Morton's also makes a variety of curing salts; we don't recommend these because they have additional ingredients in them and we prefer to add our own.) To reiterate, all salts are composed primarily of sodium chloride and that is what works its magic on meat, so any kind of salt will work.

**The Way Salt Works**

The salt in our body helps to regulate fluid exchange at the cellular level. It works this way: heightened salt concentration outside a cell (more specifically, electrically charged sodium ions) results in a fluid exchange out of the cell, water moving through the semipermeable membrane, to reduce the sodium concentration outside the cell and raise the potassium concentration within the cell. This is how our cells are fed and nourished.

The key is the semipermeable membrane, which allows certain kinds of molecules to flow through it, such as water and salt and other electrolytes, but not bigger molecules, such as proteins. A semipermeable membrane is like a tea bag, which lets flavor molecules but not the tea leaves pass through it.

When we put a piece of meat into an environment in which the salt concentration is very high, the same type of exchange happens in the cells of the meat as in our body: attracted by sodium's ions, water rushes out of the cells to join them. Equilibrium is always sought, so there is a continual back-and-forth movement through the membrane as the concentrations shift and salt in solution enters the cells of the meat (bringing some flavoring) and returns to the brine or cure (along with blood). The ionic charges also change the shape of the cells, loosening them and enabling them to contain more moisture.

By pulling water out of the meat, salt, by definition, dehydrates it. When it enters the cells of the meat, it also dehydrates the microbes that cause decay and spoilage and other potentially hazardous bacteria, killing them or inhibiting their ability to multiply. This dehydration is salt's main preservative mechanism. A secondary preservative effect is that salt reduces the water activity microbes need to live on in a piece of meat or fish.

Virtually every food group can be salted to excellent effect. Most fish and just about every kind of meat can be preserved through dry curing because of the chemical response of their proteins to salt—salt reduces the water content of the meat and creates an inhospitable environment for the bacteria that cause rot.

## Salt: Measuring by Weight or by Volume

Our salt of choice is Trapani sea salt, a medium coarse sea salt produced by Sosalt SpA in Sicily. Because salt is such a powerful ingredient, and because these recipes rely on it not only for flavor but also for safety, we only recommend salt measured by weight. If you measure by weight, you can use any kosher or sea salt you wish (avoid rock salt). If you do not want to invest in a kitchen scale but would like to use the recipes in this book, we recommend that you use Morton's kosher salt, which has a near-equal volume-to-weight ratio, meaning 1 tablespoon of salt equals ½ ounce. Thus, a recipe calling for 5 pounds of meat will require ¼ cup of Morton's kosher salt (2 ounces, or 2.5% by weight).

### Curing Salts: Sodium Nitrate and Sodium Nitrite

Nitrates and nitrites are naturally occurring chemicals that our bodies rely on. Green vegetables such as spinach and celery are loaded with them. As much as 95 percent of the nitrates in our bodies comes from vegetables. Our bodies naturally convert nitrates into nitrites, which work as a powerful antibacterial agent, particularly in an acidic environment (such as our stomachs).

Nitrates and nitrites provide that same antibacterial function in a sausage that's being dry cured. *Lactobacillus* bacteria, feeding on sugars in the sausage, produce acid (that is the tanginess we associate with salami) and the nitrites keep harmful microbes from growing—most importantly, the bacterium that generates the deadly botulism toxin.

What's rarely noted, however, is the powerful impact nitrites have on the flavor of the meat. They are what makes bacon taste like bacon, not spareribs, and what makes ham taste like ham, not roast pork. They give corned beef and pastrami their distinctive piquancy. Nitrites also are why corned beef and bacon are pink rather than gray-brown, but it's the flavor it gives to pork that, to our minds, is its more important effect.

In the 1970s, concern arose that nitrites could be carcinogenic. Recent studies have concluded that massive quantities (as in contaminated water) can do serious damage but that the amounts added to food will not. Indeed, as noted in one study, "Since 93 percent of ingested nitrite comes from normal metabolic sources, if nitrite caused cancers or was a reproductive toxicant, it would imply that humans have a major design flaw."*

And the American Medical Association reported that as of 2004, "Given the current FDA and USDA regulations on the use of nitrites, the risk of developing cancer as a result of consumption of nitrites-containing food is negligible."† Indeed, some current studies describe the health benefits of nitrates and nitrites, especially cardiovascular benefits for heart attack patients.‡

---

* Douglas L. Archer, "Evidence that Ingested Nitrate and Nitrite Are Beneficial to Health," *Journal of Food Protection* 65, no. 5 (2002): 872–875.

† www.ama-assn.org/ama/no-index/about-ama/13661.shtml.

‡ Norman Hord, "Food Sources of Nitrates and Nitrites: The Physiologic Context for Potential Health Benefits," *The American Journal of Clinical Nutrition* 90 (July 2009): 1–10.

We recommend using only what's necessary for a safe and consistent end result, not loading food up with sodium nitrite, but we don't fear it. We simply don't ingest that much of it in the form of cured meats—again, we get almost all our nitrites via vegetables.

This also means that companies advertising their products as "nitrite-free" are either uninformed or are pandering to America's ignorance about what is healthy and what is harmful in our foods. The "nitrite-free" tag is a marketing device, not an actual health benefit.

Sodium nitrite and sodium nitrate are sold under various names, with sodium nitrates including the number 2. We refer to them generically as pink salt and pink salt #2. Pink salt contains 6.25% nitrite; pink salt #2 contains 5.67% nitrite and 3.63% nitrate. Pink salt #2 is used only for long-term dry-cured sausage, or salami. The reason for this is that it contains both nitrites and nitrates; the nitrates convert over time into nitrites, so the salt acts as a kind of time-release nitrite capsule, further protection against the bacterium that causes botulism. (For centuries, potassium nitrate, called saltpeter, was used for dry curing, and it's still used in Europe. But sodium nitrite and sodium nitrate give the best and most consistent results. We don't advise using potassium nitrate.)

We recommend Butcher & Packer, based in Michigan (www.butcher-packer.com), as the best source for curing salts (and other sausage-making ingredients and equipment as well). They sell their curing salts under the name DQ Curing Salt and DQ Curing Salt #2, which is used exclusively for dry curing sausages.

Sodium nitrite is colored pink to prevent accidental ingestion: consumed in large doses, it can be fatal. The actual toxicity is 71 milligrams per kilogram: if you weighed 100 kilograms (220 pounds), ingesting 7100 milligrams, or 7.1 grams (about a teaspoon) would be dangerous.* So, while pink salt is not dangerous when used properly, it should be clearly marked and stored carefully. It keeps indefinitely at room temperature.

---

*From Oxford University's chemistry department: http://msds.chem.ox.ac.uk/SO/sodium_nitrite.html.

## Bacteria

"Can't live with 'em, can't live without 'em," says Harold McGee, the author and food-science authority. Some bacteria can kill us, but we couldn't survive without others. There are tens of millions of bacteria in a tablespoon of soil, in an ounce or two of water. Bacteria are everywhere around us. Our skin, mouths, noses, and guts teem with all kinds of bacteria. The vast majority are rendered harmless by the inhospitable environment of our bodies, and others that could be harmful are taken care of by our immune system. And some bacteria keep us healthy and aid in the digestion and the metabolizing of food.

Food doesn't spoil all on its own; it's the microorganisms—bacteria, yeasts, and molds—on its surface that turn the leftovers in the fridge from edible to inedible.

Some bacteria are not harmful to other creatures but are very dangerous to us. A relatively new strain of *E. coli*, 0157:H7, has little effect within the digestive systems of the cows where they can grow, but they can shut down our kidneys and even kill us. Other food-borne bacteria that we need to guard against are *Salmonella, Listeria,* and *Clostridium botulinum*. In the craft of dry curing, our biggest concern is *Clostridium botulinum*, the bacterium that causes botulism. It can thrive in the anaerobic (oxygen-free) environment of the interior of a sausage at warmer temperatures, the temperatures at which we dry sausages. The word *botulism* derives from the Latin word for sausage for a reason.

Under the proper conditions, botulism is not a problem, in part thanks to, that's right, other bacteria that also thrive in the interior of a sausage, such as *Lactobacillus*. These bacteria feed on sugars in the meat mixture and release an acid, lactic acid, that *Clostridium botulinum* doesn't like. Given the proper acidity and salinity, the botulism bacteria won't grow. But the salt in the mixture may not be uniformly distributed, and the development of acidity takes time. When sodium nitrite, a compound that is especially effective in an acidic environment, is also present, *Clostridium botulinum* doesn't stand a chance.

The lactic acid also has another important function: it changes the sausage's texture by coagulating the meat proteins, just as cooking would.

And, finally, the acid will give the sausage a pleasantly tangy flavor.

Thus, we need three things in our sausage mixture to ensure that it's safe and has a good texture and flavor: salt and curing salt #2, which we add, and acid, which bacteria generate.

Before they learned to cultivate bacteria, salumièri and charcutiers relied on the microbes on the surface of the meat to do their work. The good bacteria are encouraged by a salty environment and the spoilage bacteria are not. When the salumière had a good culture going, he would set aside some of the raw ground meat to add to the next batch, a practice sometimes referred to as backslopping. Bread bakers who set aside some of their yeasty dough to inoculate the next day's batch of bread dough, known as the *biga*, were doing the same thing. Relying on bacteria on the meat and backslopping, however, resulted in an inconsistent product; the sausage maker could never be sure if he had a good working bacterial culture. Today we can add good bacteria to our sausage and know with some degree of certainty how long the bacteria will take to generate enough acid to lower the pH below 4.9, the desired safety level for dry-cured sausage.

We recommend a product, Bactoferm F-RM-52, manufactured by Chr. Hansen, a company based in Denmark that specializes in what they call "natural ingredient solutions"—meaning bacteria, enzymes, and molds. They are sold in the United States through Butcher & Packer (see Sources, page 267).

Commercial bacteria, however, result in a distinct tanginess that some aficionados do not like. One maker of salumi in America, Salumeria Biellese in New York City, does not use commercial bacteria. Relying solely on developing the bacteria already present on the meat results in a subtler tang and a more traditional flavor. But it's also riskier in that, unless you have a pH meter, you have less assurance that your bacteria is working (we describe Biellese's method on pages 67–69).

Bacteria are single-cell organisms that behave like we do in that they consume food, produce waste, and reproduce. (In fact, all the beneficial microorganisms that help us make great food—bacteria, yeasts, and molds—are alive, eat, and reproduce. Viruses, often thought of in the same way we think of bacteria, are not alive; they're more like a toxin or poison, which is what *virus* means in Latin.) The bacteria that we add to our sausage grind for salami feed on the sugar, release acid, and reproduce. The colder the meat is, the slower their activity. They are at their most active at temperatures, between 85 degrees F./30 degrees C. and 115 degrees F./46 degrees C., depending on the bacterium, and work quickly to lower the pH of the sausage (that is, make it more acidic).

Bacteria feed on sugar, and the simpler that sugar is, the easier it is for them to convert it to energy. Glucose, also called dextrose (this is what we call it in the recipes), is a simple sugar and thus easy for the bacteria to consume. Table sugar, or sucrose, is a composite of glucose and fructose molecules. This means that it is less easily converted to energy by bacteria—but they can still do it. In our experience, we've found little difference in the finished product no matter what sugar we use. So, although we call for dextrose in the recipes, you can substitute granulated sugar.

Commercial starters are composed of freeze-dried bacteria and filler. The instructions say that one package (1.5 ounces/42 grams) will inoculate 500 pounds/225 kilograms of meat. In order to inoculate small amounts of meat, though, you need to ensure you get enough bacteria, and not just filler, into your ground meat. Therefore, you should add at least a quarter of a packet (at least 1 tablespoon, or 0.4 ounce/10 grams) no matter how little sausage you're making. Because the bacteria are helpful, not harmful, there's no danger in having too many. All our recipes for salami call for adding this amount of a commercial bacterial culture.

**Wild Bacteria: The Sourdough Starter of the Meat World**

With the exception of Biellese, all the salumi makers in the United States (or as far as we know) add some form of commercial starter culture to the meats they grind for salami (as well as some form of nitrate). But making salumi extends back to ancient history. Why can't we cure salami the way Italians cured salami thousands of years before commercial cultures and the Internet? We can, but few do.

Marc Buzzio of Salumeria Biellese is the only professional we know of who does not use a commercial product for his (USDA-approved) salami. He welcomed us into his production facility, showed us the curing, grinding, and drying areas. It was only after continued questioning about the process he uses to create what we think is the finest salumi in America that he and his brother-in-law and partner, Paul Valetutti, brought out a large container of meat wrapped in opaque plastic. They unsealed the bag to reveal a grayish layer of ground pork. Digging below the surface, they showed us vivid pink meat. The aroma of cured meat wafted up out of the bag. It was beautiful.

And it convinced us that the bacterial culture was well developed here during its four week "laydown," as Buzzio calls it.

But he does not believe this is where the flavor comes from; he insists that the microorganisms in his drying chamber are ultimately the reason for the uniqueness of his salami. "I'm not a scientist," he wrote us, "but I've learned over the years that it's not using a starter culture and not backbatching [or backslopping, as described on page 65] it's the room that plays the major role. When we moved to our new location, the first batches of my salami tasted like everyone else's. I had to bring them back to NYC [the older location] to restart the mold. It has taken almost three years for my new room to produce the same results. . . . Again, if my room is not right my salami aren't either, so I think it's the room."

That ran counter to our intuition, which is that when the bacteria on the surface of the meat are mixed throughout as the meat is ground and paddled, given appropriate salinity, they ferment (the same way a natural pickle becomes acidic from bacteria), and the resulting acidity prevents harmful bacteria and creates good flavor.

We queried Harold McGee, the food scientist, who said, "You're right, the acidification is going to come from bacteria mixed into the meat interior. Most microbes are not that mobile, especially in a solid medium. The stuff that grows on the surface does contribute some flavor, but it's the flavor chemicals that migrate to some extent into the meat."

A response that seems to allow for both Buzzio and us to be right.

But we also agree with Paul Bertolli, of the aforementioned Fra' Mani Handcrafted Foods, who says, "To safely make fermented salame, the pH of the meat needs to be reduced in a specified time to avoid the potential for pathogen growth. Using no starter or backslopping is risky in this respect."

For safety reasons, then, we recommend using a commercial starter so that the pH of the meat drops to 4.9 or below. That said, we think that the unique flavor Buzzio achieves is the result primarily of his natural fermentation process. We find that salami fermented with commercial cultures develop a tanginess that is slightly overbearing. But, then, it's noticeable to us because we've been tasting and making salami for so long. We believe the need for a consistent and safe product outweighs a relatively minor flavor compromise.

If you feel comfortable experimenting with natural bacteria, though, and are aware of the risks involved—and if you feel comfortable evaluating a finished salami on your own—this is what we suggest for making salami without using an added commercial bacteria starter:

- Follow the basic recipe for salami (page 120). Use only naturally raised or organic meat.
- Use the recommended level of sodium nitrate, or DQ Curing Salt #2 (see Sources, page 267), to prevent the possibility of botulism.
- Weigh the salt to ensure you have the right amount. We believe 2.75% sea salt and 0.25% sodium nitrate relative to the weight of the meat to be ideal for dry curing.
- Dice the meat and toss it with the salt, sodium nitrate, and any other seasonings you're adding. Grind it and paddle it.
- Pack the mixture tightly into a nonreactive container and cover it with plastic wrap, pressing the wrap down onto the surface of the meat.
- Put the container into an opaque plastic bag, to prevent light from turning the fat rancid. Refrigerate for 4 weeks.

• After 4 weeks, the meat will be gray (oxidized) on the surface, but underneath it should be a vivid appealing pink and have a porky, salami-like aroma.

• Repaddle the meat and stuff and dry the sausages.

As always, evaluate the finished sausage by sight, touch, and smell. It should look and smell like a well-cured product; it should not look raw or have any off aromas or scary-looking mold.

## Molds

Molds are fungi. In the world of dry curing, there are good molds and bad molds. The good molds are white and chalky, usually *Penicillium* molds. Beneficial molds perform three important functions for your salami:

• They prevent bad mold from growing by taking up real estate on the salami surface; bad mold is less likely to grow where good mold is growing.

• They consume oxygen near the surface of the salami, which retards oxidation.

• They protect the fat at the surface of the salami from light, which can make fat rancid.

Good molds can also contribute somewhat to the flavor of a sausage.

Bad molds can be various colors and textures: green, yellow, black, fuzzy or sticky. Any time you see bad mold (anything not white and chalky) growing on salumi in your drying chamber, rinse it off with vinegar and then rehang the meat. This usually takes care of the problem, though the mold can return. Repeated problems can require washing down all the surfaces of your drying chamber with a bleach solution. Some molds can burrow through casings and work their way into a sausage's interior if left untended, which would require throwing the sausage out.

Initially salumièri relied on the microflora present in the dry-curing room landing on the hanging sausages and growing there. If they had problems developing good molds, they hung sausages with good mold already on them next to newly hung salami to encourage the spread of the good mold. Today commercial mold cultures are available and work well. We recommend Mold 600 Bactoferm, made by Chr. Hansen and sold here by Butcher & Packer (see Sources, page 267). If white molds don't grow naturally in your environment, the commercial cultures are worth using. They protect the sausages against all their external enemies—bad molds, harmful bacteria, oxygen, and light. You shouldn't need this for whole muscles, but if you've got a serious mold problem, it can be used on them.

To use a commercial mold culture, dissolve it in distilled water and spray or brush it onto the sausage, or dip the sausage in the solution; follow the instructions on the package. The cultures are highly recommended for home sausage curing and for developing good mold generally in a dry-curing room (though it's optional in all the recipes). Brian needed to inoculate one sausage years ago, and his whole drying room has been filled with benevolent microflora ever since.

## Other Equipment and Materials

As mentioned earlier, if you intend to break down whole or half animals, it's useful to have a meat saw. Here we discuss other recommended equipment.

### Scales

A digital scale is one of the most important tools in your kitchen. Good ones start at around $30 and are worth it if you're making anything in this book. If you don't have one, you are handicapped in your ability to make, in addition to so many other things, good consistent salumi. If you don't have a scale, there are ways to fudge, but your results won't be consistent (see page 80 for advice).

Recipes are really relationships between ingredients, or percentages of one ingredient relative to another, and the only reliable way to measure this relationship is by weight. For instance, to make a perfectly seasoned, well-cured salami, you need to add between 2.5 and 3% of the meat mixture's weight of salt. So, if you were making a 100-ounce batch of salami (a little more than 6 pounds), you'd add 2.5 to 3 ounces of salt; if you were making a 1000-gram batch, you'd need to add 25 to 30 grams. Because the weight of a tablespoon of salt varies so much from one salt to another, you won't know how much salt you're actually adding to the meat unless you measure it by weight.

We use two kinds of scales, one that goes from a single gram/fraction of an ounce to more than 2250 grams/5 pounds, and one that measures small amounts in tenths of grams. (Brian's students call this his drug-dealer scale.) For recommendations, see Sources, page 267.

### Meat Grinders

If you grind a lot of meat, it's useful to have an electric meat grinder. For small amounts, and for homemade salumi, grinder attachments for standing mixers or old-fashioned hand-crank grinders are OK, as long as the blades are sharp, but we don't really recommend them. Too often the blades and dies are not sharp, so they heat up

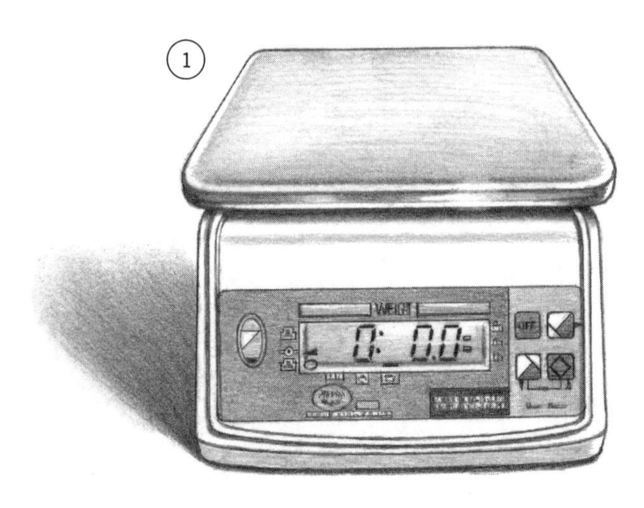


1. A digital scale is a critical kitchen tool for measuring your ingredients accurately.

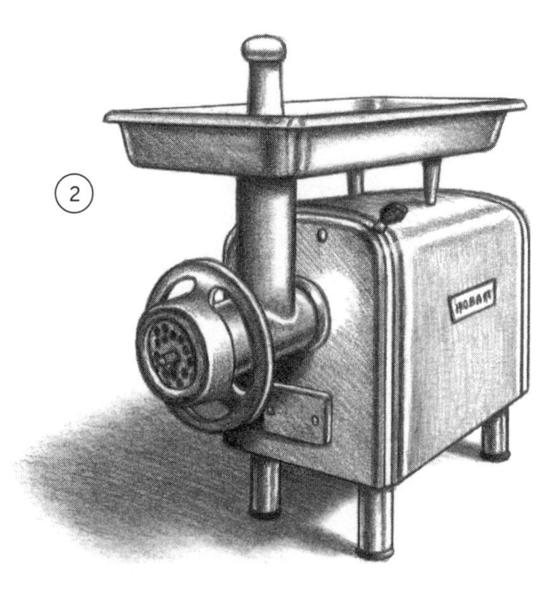


2. A professional meat grinder has a larger capacity and more power than a home grinder. The better the grind, the better the sausage.

3. There are many models of home grinders available; choose one that matches the amount of production you plan on. (Grinder attachments for stand mixers are acceptable for occasional grinding, but we don't recommend them for serious sausage making.)

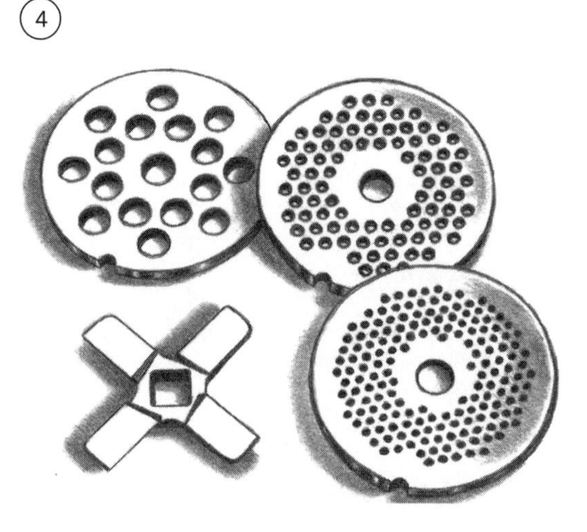


4. Grinder dies and cutting knife. The size of grind is important in sausage making. The three most common die sizes are ⅛ inch (3 millimeters), ¼ inch (6 millimeters) and ⅜ inch (9 millimeters).

the meat with too much friction, which can give the sausage an off texture. So, we recommend a stand-alone meat grinder if you're going to get serious about making sausage (see Sources, page 267). As a rule, the faster and cleaner the cutting, the better the sausage will be. These meat grinders also come with heavy-gauge dies in multiple sizes, which gives you more control over the texture of your sausage.

### Casings

Casings to contain the meat can be anything from sheep casings for small breakfast sausages, to hog bungs, beef middles, hog bladders, or laminated (machine-made) casings of varying sizes. We've tasted the salami of one farmer–salumi maker in Tuscany who cured his meat in brown butcher paper. The only requirement is that it be porous to allow moisture out.

Traditional salami is made using larger casings, hog bungs or beef middles. The smaller they are, the easier and quicker the dry curing will be. For all our salami, we recommend using beef middles, which are about 2.5 inches/6.5 centimeters in diameter, cut into 18-inch/45-centimeter lengths. These are not so big as to make dry curing difficult, but they are big enough to provide a satisfying slice. However, any size casing can be used to cure salami (see Sources, page 267).

All casings should be soaked in water for at least 20 minutes, and up to 24 hours. They should then be flushed with water before you stuff them.

We recommend peeling off the casing before eating salami. (It's what they do in Italy.) The casing is edible but it can be tough and is often covered with mold.

Most casings are sold packed in salt. Unused casings can be returned to the salt and refrigerated for at least a year. Be sure to add more salt so that they are completely packed in it. We've had some casings for five years that are perfectly fine. If they become moldy or slimy or have a bad odor, throw them away.

### Sausage Stuffers

The most convenient and easiest stuffers are the cylindrical ones with a hand crank that lowers a plunger. If you make a lot of sausage, these are worth the expense (starting at $130). We recommend a stuffer that holds 5 pounds (see Sources, page 267). The

## PREPPING CASINGS BEFORE STUFFING

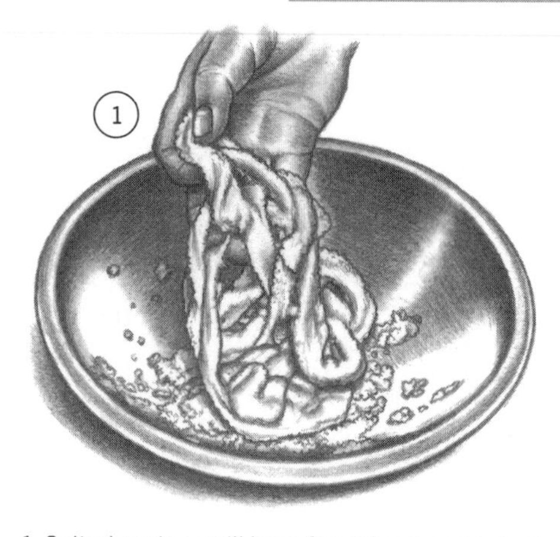


1. Salted casings will keep for at least a year in the refrigerator. Before using, they should be soaked and flushed with water. First, shake off the excess salt.

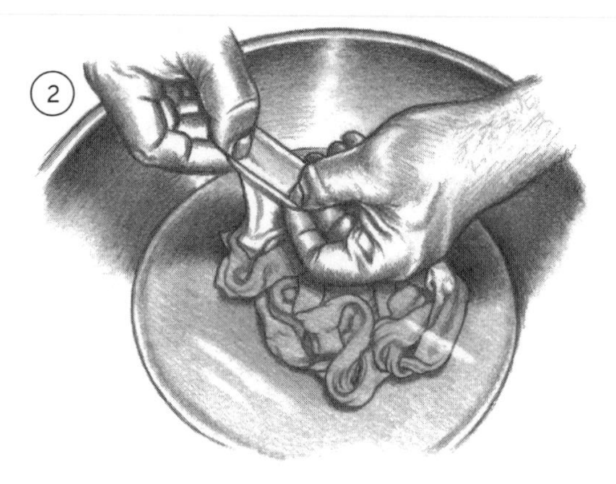


2. Rinse the casings in several changes of water to remove any remaining salt. Hold one end of each casing under the faucet and flush with cold water.

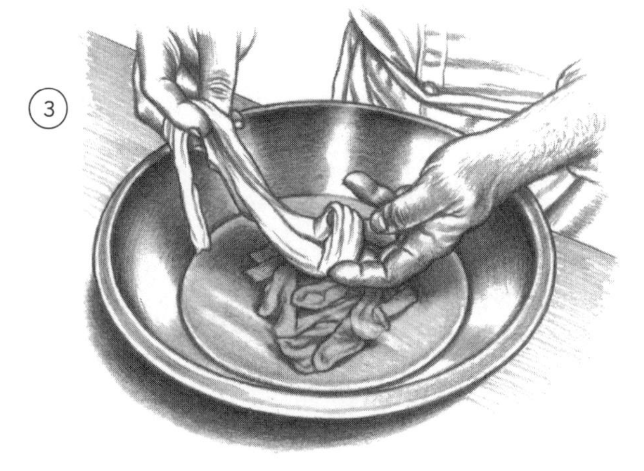


3. Stored in cold water, the rinsed casings will keep for 3 to 4 days in the refrigerator. Never freeze natural casings.

## STUFFING EQUIPMENT

1. A cylindrical plunger sausage stuffer is the best tool for stuffing sausage. Make sure the ground meat is packed tightly in the hopper, without any air pockets.

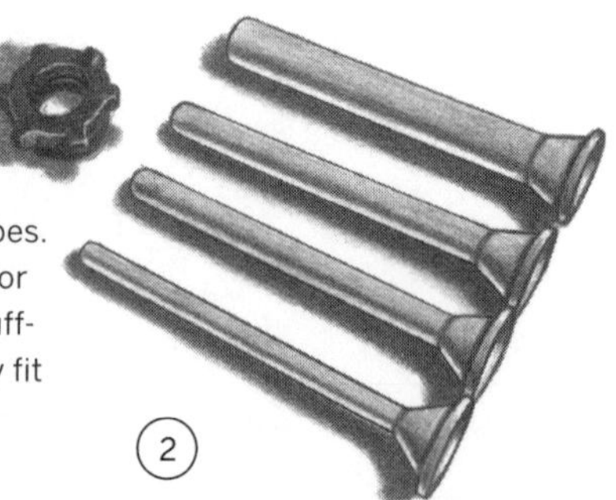


2. Sausage-stuffing horns, or tubes. Choose one that is appropriate for the size of the casing you are stuffing; the casing should just barely fit around the horn.

attachments for standing mixers work but are messy and difficult to use, and they don't result in the best texture; they are acceptable if making sausage is only an annual event. If you're making small amounts of large dry-cured sausages using beef middles or beef bungs, stuffing these big casings by hand is not difficult.

**Smoking Devices**

Some recipes here call for cold-smoking. Cold-smoking is defined by a temperature of less than 90 degrees F./32 degrees C. or lower so that the meat is smoked but not cooked. Heat is necessary to create the smoke; therefore, the heat source for the smoke has to be separate from the chamber containing the meat. Without the proper equipment, cold-smoking can be difficult. There are few products on the market that can be used in a single chamber (e.g., a kettle grill), such as the smoke gun by Polyscience and the A-MAZE-N-SMOKER, which give foods smoke with little heat. Or you can jerry-rig a cold smoker in a kettle grill or build a makeshift smoker using a large cardboard box and some clothes-dryer tubing through which to pump smoke. There are lots of how-to's on the Internet if you're the tinkering type.

Regardless of whether you have a proper cold-smoking unit or have fashioned one of your own design, the conditions you need are the same: an enclosed chamber for the meat and smoke and a temperature below 90 degrees F./32 degrees C. Recipes suggest how long the meat should stay in the smoke for optimal results, but you can vary these times as dictated by your equipment. Use your common sense and your sense of taste to guide you.

Some recipes here call for hot-smoking. Hot-smoking is defined by a temperature of 200 degrees F./93 degrees C. Inexpensive stovetop smokers are available, as are more pricey outdoor models. Hot-smoking both smokes an item and cooks it, but at a low enough temperature that it cooks gently. (We consider smoking at temperatures of 300 degrees F./150 degrees C. to be smoke-roasting.)

Jerry-rigging a hot-smoking device is easier than setting up a cold-smoking device. It can be as simple as building a very low charcoal fire in a kettle grill and covering it with soaked wood chips. Another strategy is to smoke the item first, then finish it in a low oven (200 degrees F./93 degrees C.).

# The Environment: Creating a Place to Dry Cure Meat

To our knowledge, there is no small appliance made specifically for dry curing, so if you want to dry cure meat and sausage, you'll have to create the proper environment on your own. Brian is lucky enough to have a walk-in cooler at Schoolcraft College, where he teaches, dedicated to dry curing. If you have enough cooler space to do the same, you'll need to be able to raise the temperature to 55 or 60 degrees F./12 to 15°C. and you'll need to keep the air humid and continuously circulating.

To those chefs who want to dry cure meats and sausages to serve at your restaurants, we can only say that local health departments rarely, if ever, have protocols or HACCP plans for the production and sale of dry-cured meats, meaning that salami and coppa fall under the same rules as raw product. (HACCP = Hazard Analysis Critical Control Points, a NASA-developed protocol for maintaining food safety—google it for more info.) And most health departments will force you to throw away meat that has been stored above 40 degrees F./4 degrees C. for more than three or four weeks, even though it's perfectly safe to eat if it has been properly cured and dried. If you want to create a dry curing program for your restaurant, we recommend educating your health department and creating your own HACCP plan (see page 26).

For those who want to cure only small amounts of meat and sausages, here are a few suggestions.

You need to create the conditions that will allow the meat and sausage to dry properly, but there is no set way to do this. The ambient humidity should be between 60 and 70 percent; the temperature should remain between 55 and 65 degrees F./12 and 18 degrees C.; and the air must circulate. If you create these conditions, you should be able to dry cure successfully. But you must also keep in mind that some of the best-known charcutiers and salumièri in the world, who have built thriving dry curing businesses are regularly confounded by inconsistencies and failures. So be prepared for anything to happen. The craft of dry curing meat relies on a wild world of microbes; some of them wear white hats, some of them wear black hats, and none of them listen to you or me. The only thing you can do is create conditions that favor the good microbes over the bad microbes and pay attention to what happens.

The first thing we do to encourage the good guys is to use the right amount of salt, which is about 3% by weight of the meat mixture. This is about the same salinity as the ocean, where conditions conducive to life allowed organisms to evolve. A 3% salinity is conducive to the *Lactobacillus* bacteria that eat sugars and generate lactic acid that makes it even more difficult for the black hats to get a stronghold. Bad molds, black or green and furry molds, can cover a sausage if the good white penicillin molds don't get there first and grow faster. Bad mold is unlikely to grow where good mold is already growing.

We have successfully cured in a miniature refrigerator, which when turned to its warmest setting will remain between 55 and 65 degrees F./12 and 18 degrees C. In order to ensure adequate humidity in the fridge, put a small pan or bowl of heavily salted water (equal parts, basically a salt slurry, so that mold doesn't grow on the water). There's one main adjustment you'll need to make. The freezer section gathers moisture that, at these temperatures, can drip down onto anything below it, so you'll need to rig some way of diverting the water. A piece of foil can be slung below it and tilted toward the back so it drips into the pan of salted water.

We have also successfully dry cured meats in wine refrigerators, also using a pan of salted water for humidity, and in wine cellars, as well as within plastic-wrapped speed racks.

If you're handy and wanted to build a dry-cure box, it wouldn't be difficult. You could fashion a wooden box and cover the insides with something nonporous, such as FRP (fiberglass-reinforced plastic) board, anything that won't absorb moisture and can be wiped down. You'll need a cooling device if you don't have a cellar or other space that remains cold, as well as a source of humidity and circulating air.

Again, there's no special equipment designed for this. Maybe there will be one day, but until then, you're left to your own devices, whether with a mini-fridge, an old refrigerator converted to give you the required temperature, a wine fridge, or some sort of construction of your own devising. No matter which it is, the same things hold true: You need:

- 55 to 65 degrees F./12 to 18 degrees C.

- 70 percent humidity

- Air circulation

That, and the common sense to evaluate how well what you're using is working.

## About the Recipes: Read This First!

All the recipes assume that you have a basic working knowledge of the kitchen and have a sixth sense that is always on, called "common." Because repeating all the myriad basics in each recipe would make them unnecessarily cumbersome, we will spell that information out here for you to refer to if necessary.

**Curing Whole Muscles** (**pages 87–115**)

Many of the recipes in the first section of this book do not give specific weights for the cuts or amounts for the other ingredients because these will be variable. You can't always get, for instance, a pork loin that weighs precisely 5 pounds/2250 grams. So, because the weight of the meat can never be standard, we give the general for most boneless whole muscle cuts (coppa, lonza, boneless ham) and use what we refer to as the "salt-box method."

Most whole muscles don't require pink salt to prevent botulism when they're dry cured. We do recommend pink salt for the rolled pancetta (page 109). Pink salt also gives pork a bacony flavor, which many like and we prefer, both in belly and in guanciale. When curing whole muscles, add pink salt, sodium nitrite, if you wish, at 0.25% of the weight of the meat (multiply the weight of the meat by 0.0025).

Whole muscles can be cured within a casing or simply tied and hung. Long, thin cuts are hung simply by poking a whole in a corner of the meat and slipping string through the hole. If it's a particularly heavy cut, you may need more than one piece of string. For long, cylindrical muscles (i.e., coppa and lonza), we tie the meat as you would a roast, using one continuous piece of string and hitches (see the illustrations on pages 122–25).

THE SALT-BOX METHOD (*recommended for all whole-muscle curing*)
Using the salt-box method simply means dredging a cut of meat in salt. Fill a container with kosher or coarse sea salt and roll your meat in it until it's uniformly

Prepare the salt mixture and place it in a suitably sized pan. Be sure to mix the spices well so they're evenly distributed throughout the salt.

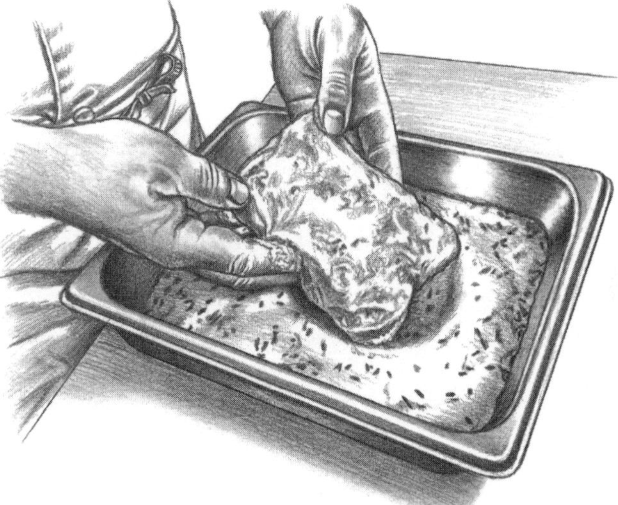

Dredge the meat in the salt mixture. With your hands, pack the salt onto the muscle, completely covering its surface. (Use only the amount of salt that naturally sticks to the meat.)

coated. That's all there is to it. This will work with all whole-muscle cuts, from guanciale to culatello. They should be uniformly coated with salt, put in a large plastic bag so that the cure salt and any liquids pulled from the meat stay in contact with it, and refrigerated for 1 day per each 2 pounds/1 kilogram of the meat's weight. If you want to add seasonings to the cure, such as black pepper, do so by eye after you've salted the meat.

Large bone-in cuts, such as hams, should be fully packed in salt, weighted down, and refrigerated for the same amount of time.

For pancetta and guanciale, we provide more precise measurements because we use additional seasonings, but you can always use 3% salt by weight (multiply the weight of the meat in ounces or grams by 0.03 salt). Most whole muscles can also be salted with 3% salt by weight, provided you allow enough time for the salt to penetrate to the center, establishing an equilibrium throughout the cut. Be aware that meat salted at 3% is pleasantly salty when eaten thinly sliced and cold. If you're cooking with it, it will taste considerably more salty and should be used in conjunction with other ingredients.

### What to Do If You Don't Have a Scale

We don't recommend making salami without a scale, but if you must, here's what to do. Have the butcher cut and weigh out the appropriate amount of meat and fat. Use Morton's kosher salt, which as a nearly equal weight-to-volume ratio (that is, 1 tablespoon weighs ½ ounce/14 grams). Measure the sodium nitrate by the teaspoonful: 1 teaspoon of pink salt weighs about ¼ ounce/7 grams.

Be aware, though, that butchers have varying sanitation practices, another reason we recommend you do your own cutting and weighing for dry-cured meat.

### Making and Curing Salami (pages 116–58)

For all sausages, both fresh and dry cured across the board (not just the ones in this book), the meat and fat should be diced to a size that can be dropped down the shoot of your grinder—that is, not so big that you have to use a plunger to stuff it down.

Most of our recipes suggest partially freezing the meat and fat before grinding. This helps to ensure that the meat and fat adhere to one another. If the meat and fat are too warm, they can break the way an emulsion breaks, and the effect is just as bad. Moreover, the act of grinding heats them up with friction. So, to make sure your meat is very, very cold, put it on a baking sheet in the freezer until it is stiff but not frozen solid, 20 or 30 minutes or so. Fat can be completely frozen, if you wish, or partially frozen, like the meat. If you are making a lot of sausage using a big metal grinder, it's also advisable to freeze your grinding tools—the feeder, auger, blade, dies, everything—before grinding.

We have used meat grinder attachments for stand mixers for salami, and they do work if that's all you have, but your meat is more likely to smear with one of these attachments. Smear—when the meat comes out of the die as a kind of pale paste—means a bad sausage. If you're serious about sausage, get a good meat grinder (see page 71 and Sources, page 267). We don't recommend hand-crank grinders for these sausages.

After sausage meat is ground, seasonings, liquids, and cultures are often mixed into the meat. This is best done using a stand mixer with the paddle attachment. The paddle distributes the meat and fat, seasonings, and other ingredients. It also develops the myosin protein in the meat, which is sticky and helps the sausage hold together, giving the sausage a good "bind." Most of our recipes instruct you to mix until the meat becomes tacky, almost furry—that's the texture you want.

For information on sausage stuffers, see page 73.

For information on casings, see page 73.

STUFFING SAUSAGE • Whack the ground meat down into the cylinder to avoid air pockets. Fit the stuffer with the largest tube available, thread the casing over it, and crank away, keeping enough tension on the casing so that the meat fills the casing tightly. Using a sterile needle or a sausage pricker, prick any air pockets.

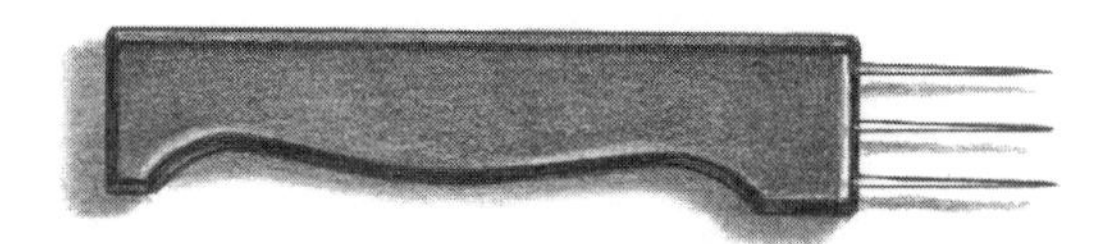

A sausage pricker is a convenient tool. If you do not pack your sausage tightly into the casing, there will be air pockets; these pockets of trapped air can result in bacterial growth. Twist both ends of the sausage and gently prick the casing to release the air. If you don't have a pricker, a needle works well.

Both ends of the sausage are tied with a bubble knot, a knot that catches some of the casing within the tie. This prevents the wet, slippery casing from slipping out of the string. See the illustrations on pages 122–25.

SEASONING • We recommend you weigh your salt to ensure accuracy. Most salami recipes call for between 2.5 and 3% salt (including curing salt). We recommend 2.75% salt and 0.25% sodium nitrate (such as DQ Curing Salt #2). These recipes call for 5 pounds of meat, but if you have more or less, calculate the amount of salt by multiplying the weight of the meat by these percentages and weighing the salts.

If you're using a skin-on cut and all the bristles have not been removed, remove them with a razor or burn them off with a torch.

**Containers and Weights**

CONTAINERS • Use common sense here. Don't cure a little piece of meat in a giant tub where much of the cure won't be in contact with the meat. Most cuts other than whole bone-in hams fit in a 2.5-gallon/9-liter zip-top plastic bag. This is the best and most convenient option. It's nonreactive to the salt, it keeps the cure close to the meat, and it makes it very easy and clean to rerub ("overhaul," in professional terms), the cure over the meat to redistribute it. If you aren't able to find these bags, use a plastic or glass container in which the meat fits snugly.

WEIGHTS • Many of these recipes require you to weight the meat you're curing. This helps to press water out of the muscle for more efficient and effective curing. Use whatever you have at hand: canned goods, cast-iron pans. A brick weighs about 4 pounds/2.2 kilograms. If a recipe tells you to weight the meat down with 8 pounds, put a tray or pan on top of the meat and put a couple of bricks or six 28-ounce cans of tomatoes on it. The exact weight is not critical.

### Herbs and Spices

We use a lot of paprika in these recipes (it's far less common in Italy). Our brand of choice is La Vera (see Sources, page 268). Use whatever brand you wish, Hungarian, Spanish, American, just make sure it's fresh. Don't use the paprika that's been sitting in the spice rack you got as a wedding gift.

USING WHOLE SPICES • It's important when using whole spices to toast them first to release their flavorful oils and expand their flavor, especially in dry curing, where the spices won't be heated and cooked (as they are when, say, cracked black pepper is used to season a steak). Black peppercorns and coriander seeds are used frequently in these recipes. Toast these in a sauté pan for a few minutes and then crush them using a mortar and pestle, or put them on a cutting board and drag a sauté pan heavily over them several times until they are well cracked.

DRIED HERBS • We often use dried herbs for their intensity of flavor, but, as with spices, don't use old store-bought herbs. In fact, we recommend you grow your own herbs if you can and dry them yourself. Thyme grows like a weed, and it can be cut back regularly throughout its growing season and kept in an open bag to air-dry. When buying dried thyme, only buy whole leaf; the oils in ground thyme are volatile and the taste tends to be insipid even when just purchased.

## Rules of Thumb:
## Salt, Time on the Cure, and Casing

In Europe, the virtually ubiquitous curing rule is that whole muscles are cured in salt for one day per kilogram. The exact amount of salt is not critical, as long as there is enough to completely coat the exterior, but the time on the salt is important; some salumi makers use only enough to give the muscle a solid coating; bone-in hams are almost always completely packed in salt and weighted. Time on the salt is what matters: one day per kilogram or every 2 pounds. (Do use your common sense here; if the meat is still so squishy after curing that you want to leave it on the salt a little longer, do so.)

After the meat has been cured, it's rinsed of any remaining salt (often with wine, because it tastes better than water, and is more romantic besides) and then prepared for hanging. Thicker muscles benefit from some kind of protective barrier. In the case of hams, front and back, the skin can provide that protection. Where the meat is exposed, often a paste of equal parts by volume of flour and lard, called *strutto*, is smeared on the exposed flesh. Larger skinless muscles—the coppa, for instance, or the culatello—are best contained in some form of casing.

In Italy, that casing is usually the bladder (prepared by being blown up like a balloon to stretch it out and dried that way, then rehydrated and sewn around the muscle. Bladders are hard to come by in the United States, at least at this time, but there are other options. The easiest one is to wrap the muscle in untreated, brown butcher paper (nonwaxed; see Sources, page 267). There are also laminated casings, bladder-shaped casings machine-made from actual casing (this is what we recommend for culatello; see page 174). Depending on the cut, you could also use large salami casing, such as cow bung, cut and sewn or tied around the muscle. It almost doesn't matter what you use so long as it's thin and porous, to allow moisture to leave the muscle. One exception:

the paletta cotto (page 171), which is dried but then simmered, so paper casing won't work. All of the recipes give a specific method or choice for enclosing the muscle, but any method you wish should suffice.

Finally, again, there are safety issues, particularly with salami, so it's important to read the section on salt and curing salts.

On reusing the cure: Don't. Always discard the cure after removing the meat from it.

---

**Knowing When the Salumi Has Dried Long Enough**

There's an easy and accurate way of determining when your meat has dried long enough: weigh it. As a general rule, meat has dried sufficiently when it has lost 30 percent of its weight. Of course, in order to determine this, you have to weigh the meat before it goes into the drying chamber. Write this weight down somewhere, ideally on a strip of painter's tape that you can affix to the string from which the meat hangs. Over time, you'll get a sense of "doneness" by touch, feeling the level of firmness and give, but until you do, use your scale.

**Serving and Keeping Salumi**

Almost all salumi should be served at room temperature, thinly sliced. The only exceptions are the smaller salami. Thickness of the slices here is completely an issue of how it eats, what the experience of chewing it is.

Leftover salumi is best stored wrapped in butcher's paper, but it can be wrapped in plastic. It will keep well for 1 to 2 weeks this way in the refrigerator but will eventually dry out and pick up fridge odors, so it's best not to leave it for too long. It will take many months for it to go bad and grow mold (at which point it should be discarded), but generally the sooner you eat it once you've cut into it, the better.

# 3. Salumi: Recipes and Techniques for the Big Eight

# Guanciale
## (JOWL)

We consider the jowl to be one of the most magical of the Big Eight cured cuts. The fact that it's from the jowl—from the face of the hog, a part of the animal we never used to think of eating in America—and results in some of the finest and most versatile salumi makes it special to us. In Italy, a cured hog jowl is called *guanciale*, meaning pillow (one of the many reasons to love this language!).

You can cure the jowl for slicing and eating, or you can cook with it. When you cook with it, it can give the food that mysterious unnameable depth that takes a dish from good to great. Brian likes his jowl very salty, and so he cures it with more than twice the usual salt for twice the time on the cure. He uses guanciale as a seasoning, as flavoring for a stew or soup, rather than a main component in a dish, so the additional salt ensures that the guanciale has a distinctive impact. If you have different uses in mind, such as slicing it and serving it like lardo, guanciale can be cured with the standard 3% salt by weight or, better, by using the salt-box method (see pages 79–80), with the standard one day per 2 pounds/1 kilogram time on the salt.

Jowls come in varying sizes and shapes, depending on the hog and the butcher. Yours may come with the cheek meat, or maybe the butcher wanted to braise this tough but delicious lump of meat for himself—in which case it won't. It's preferable to get it with cheek meat attached, but the cure and use of the guanciale remains the same.

Be sure to check the jowl for glands, which are common from the back of the head and into the shoulder. They feel kind of like lumps of too-soft fat, but they're discolored, a kind of grayish, off color, distinct from meat or fat, and should be removed.

If you want to cure the jowl for slicing thin and serving, or for *spuma di gota* (see page 221), which is whipped guanciale, leave it to dry for 5 weeks. If you'll be cooking with it, it will be ready to use in about 3 weeks. The drying time is less important when you'll be cooking it; the longer it dries, the more flavor it develops.

Guanciale is commonly used in pasta dishes, diced or sliced and sautéed to render its fat and flavor the pasta sauce (see the carbonara on page 234). It can also be rendered to begin the base of a soup or stew or tomato sauce; add onions and aromatics to the rendered fat to cook them before adding your main ingredients. It can also be diced and folded into sausage or pâté mixtures as a garnish. (Among the best uses for fresh jowl, by the way, are in pâtés, terrines, and sausages; add it in place of fresh back fat. It suspends well in the protein, resulting in great texture and adding deep flavor to these fatty preparations.)

You can use sodium nitrite here if you wish, for a more bacony, pancetta-like flavor, using 0.25% of the weight of the meat (that is, multiply the weight of the meat by 0.0025 to determine the weight of the sodium nitrite; about ¼ ounce for 5 pounds, or 6 grams for 2 kilograms, of meat. But we mostly keep our whole-muscle recipes simple with salt and seasonings, since bacterial contamination inside the meat isn't an issue.

# Guanciale with Black Pepper

A simple cure maintains the essence of this special cut. But the jowl is very similar to the belly, so if you'd like to experiment, use aromatics commonly used for pancetta, such as garlic and thyme.

**THE CURE**

**2 ounces/60 grams sea salt (salt-box method, see pages 79–80, or 3% of the weight of the jowl)**

**2 ounces/56 grams black peppercorns, roughly cracked in a mortar with a pestle or beneath a sauté pan**

**One 4.5-pound/2000-gram hog jowl, skin on, all glands removed**

**Dry white wine for rinsing the meat (optional)**

**1.** Combine the salt with 1 ounce/28 grams of the black pepper (it should be roughly but well cracked, so that it coats the guanciale evenly). Slip the jowl into a 2.5-gallon/9-liter zip-top plastic bag and pour in the cure. Rub the jowl well with the salt and pepper. Seal the bag, squeezing out as much air as possible.

**2.** Put the jowl on a baking sheet, put another pan on the jowl, and weight it down with about 8 pounds/3600 grams of weights. Refrigerate for 2 days.

**3.** Rub the salt and juices around the jowl to redistribute them. Flip the jowl and reweight it. Refrigerate for 2 more days.

**4.** Remove the jowl from the bag and rinse it under cold water, then pat dry. Rub with the white wine if you wish. Sprinkle the jowl with the remaining 1 ounce/28 grams cracked pepper. Poke a hole through a corner of the jowl, run a piece of butcher's string through it, and knot it. Hang the jowl in the drying chamber for 3 to 5 weeks, or until it has lost 30 percent of its weight.

**Yield: One 3-pound/1300-gram guanciale**

# Coppa

## (NECK/SHOULDER/LOIN)

The coppa is striated with several distinct lines of fat, almost dividing this round muscle into quadrants. It's the fat combined with the flavorful meat of this moderately worked muscle and the ease with which it cures that makes it perhaps our favorite cut to transform into salumi. It's unquestionably the best choice for the home cook with limited dry-cure storage or environmental control to dry cure for salumi. Pancetta rivals it for flavor and ease, but sliced thin and eaten as is, nothing beats it. Coppa is especially versatile in terms of the different flavor directions you can take it in, whether floral, using fennel or a similar aromatic seasoning, or spicy, using dried hot chiles. Regardless of how it's flavored, coppa is best thinly sliced.

Isolating the actual coppa is critical. It's the muscle starting right behind the ear, against the spine, and running above the first six ribs, where it begins to segue into the loin (see illustrations 1 and 2 on page 39, and illustration 3 on page 39, which shows it removed from the spine-rib section; note that in the American hog breakdown, the cut is between the third and fourth ribs, cutting the muscle in half). When you're buying coppa, especially in Italy, be aware that it may also be referred to as capocollo or filetto (the name we give to the tenderloin).

## Coppa with Black Pepper

Coppa benefits from some seasoning. Here simple black pepper suffices (we use about a tablespoon for every 5 pounds/2250 grams or so in the cure and twice that for the aromatic coating). If you have meat from a quality hog, that's all you need.

**Coppa**

**THE CURE**
**Coarse sea salt or kosher salt**
**Black peppercorns, toasted and roughly cracked in a mortar with a pestle or beneath a sauté pan**

**Dry white wine for rinsing the meat (optional)**
**Black peppercorns, toasted and finely ground**

**1.** Weigh the coppa, dredge it in salt (see The Salt-Box Method, pages 79–80), and put it in a 2.5-gallon/9-liter zip-top plastic bag. Add peppercorns to the bag, 1 or 2 tablespoons, or until it looks good. Mark the bag with the coppa's weight and the date. Squeeze as much air out of the bag as possible and seal the bag.

**2.** Put the coppa on a baking sheet. Put another pan on top of the coppa and weight it down with 8 pounds/3600 grams of weights. Refrigerate for 1 day per each 2 pounds/1000 grams. Midway through the curing, flip the coppa redistributing the salt and pepper as you do so, and weight it again.

**3.** Remove the coppa from the bag and rinse it under cold water. Pat dry with paper towels, and rub with the wine if you wish. Weigh the meat if you intend to determine doneness by weight. Dust with finely ground pepper, evenly coating all surfaces.

**4.** Tie the coppa as you would a roast (see the illustrations on pages 122–25) and hang in the drying chamber for 4 to 6 weeks, or until it has lost 30 percent of its weight.

**Yield: 1 coppa**

# Spalla
## (SHOULDER)

The spalla is the outer shoulder muscle of the hog, and it cures just as all other muscles of the hog cure. All of the cured larger muscles have a prosciutto-like flavor, but because the shoulder cures faster than the massive back ham, it will never have the depth of funky, nutty, savory sweetness of a great prosciutto. But it's a much easier cut to cure, and a faster one too. It's more a cross between *lonza*, the lean loin, and prosciutto.

Its shape has a big effect on what you can do with it. The shoulder is a versatile, mutable grouping of muscles. It can be cured on the bone or boned and butterflied. Leaving the shank in makes it easier to hang and requires a longer drying time, enhancing the flavor. Taking the bone out speeds the drying. What you do with it depends on your situation. (If you do cure it on the bone, be sure to save the bone for soup; it's especially good in bean soups.) The shoulder can also be boned-out, salt-cured, and rolled and stuffed into a large casing to hang and dry. Or it can be cooked. One of the most intriguing preparations we discovered in Italy, in a little mountain town outside Biella, was a shoulder that was cured and then cooked (see page 170). And you can take spalla in any flavor direction you want—plain with just pepper, savory with lots of herbs, or spicy.

Always trim off any loose fat or meat. If using a boned shoulder, be careful when shaping it—tightly rolling and tying it, for instance—to make sure that there are no trapped air pockets where mold could grow.

The following are recipes for the two primary preparations of dry-cured shoulder, on the bone and off the bone.

# Air-Dried Bone-In Spalla

This is probably the easiest way to cure the whole shoulder. It gives you the same satisfaction that prosciutto offers in considerably less time. It's a great starter recipe for curing a whole bone-in cut. We recommend you remove the shoulder blade (see illustrations 1 and 2 on page 39) for easy carving. Treat it as you would a prosciutto, sliced thinly and served as is, or on a salad, or with melon. You can julienne or dice it and use it as a seasoning wherever you might use bacon—in pasta sauces, or to start a sauce or a soup. When using it this way, as when cooking with any salt-cured meat, season your dish carefully, being aware that the salt in the meat will be drawn out into the liquid.

**THE CURE**

**19 ounces /535 grams sea salt (salt-box method, see pages 79–80, or 8.5% of the weight of the shoulder)**

**2 ounces/56 grams black peppercorns, toasted and roughly cracked in a mortar with a pestle or beneath a sauté pan**

**One 14-pound/6300-gram bone-in pork shoulder, shoulder blade and arm bone intact**

**Dry white wine for rinsing the meat (optional)**

**1.** Combine the salt and pepper in a 13-gallon/49-liter plastic bag and mix well. Put in the spalla and rub the cure thoroughly all over the meat. Seal the bag.

**2.** Put the shoulder on a baking sheet, top with another pan, and put 8 pounds/3600 grams of weights on it. Refrigerate for 6 or 7 days. Halfway through the curing, rerub the shoulder with the salt and juices to redistribute it, flip it, and weight it again.

**3.** Remove the shoulder from the bag and rinse it under cold water. Pat dry. Rub with the wine if you wish.

**4.** Tie a rope around the end of the shank and hang in the drying chamber for 7 to 10 months, or until it's lost 30 percent of its weight.

**Yield: One 10-pound/4500-gram spalla**

## Air-Dried Boned Spalla

We tasted excellent boneless spalla at Salumeria Sergio Falaschi in San Miniato, a picturesque Tuscan town on a hill (there must be at least one Tuscan town that isn't picturesque, yes?), and it's an excellent way to cure this cut. When you remove the bone, you are essentially butterflying the cut, increasing its surface area, which will allow it to give up its moisture more quickly than if left on the bone—thus decreasing the drying time; the time on the cure remains the same. And, of course, it's easier to slice.

We use the salt-box method here. You'll need about 2 pounds/1000 grams of salt and a couple of ounces, about 60 grams, of pepper per pound of salt.

**THE CURE**

**Coarse sea salt or kosher salt**

**Black peppercorns, toasted and roughly cracked in a mortar with a pestle or beneath a sauté pan**

**1 bone-in pork shoulder**

**Dry white wine for rinsing the meat (optional)**

**1.** To bone the shoulder, lay it skin side down on the work surface. Locate the joint where the shoulder blade and arm bone connect, and separate the joint. Make an incision following the shoulder blade on both sides. Detach the shoulder blade from the arm bone joint cut around the tip and pull back in one sweeping motion to remove the shoulder blade (see the illustrations on page 39). Following the natural seam, cut along both sides of the arm bone and remove the bone with the shank intact. Trim the meat or fat to ensure the muscle is relatively even in thickness.

**2.** Put enough salt in a large roasting pan or container to encase the shoulder. Put the meat in the cure and coat with salt, scooping salt over the top of it and turning it and pressing it into the salt so that it's uniformly covered (see The Salt-Box Method, pages

79–80). Put it in a 13-gallon/49-liter plastic bag and seal the bag. If it won't fit, cure the shoulder in a nonreactive vessel.

**3.** Put the shoulder on a baking sheet, top with another pan, and put 8 pounds/3600 grams of weights on it. Refrigerate for 1 day per each 2 pounds/1000 grams. Halfway through the curing, rub the shoulder with the salt to redistribute it, flip it, and weight it again.

**4.** Remove the shoulder from the container and rinse it under cold water. Pat it dry, then rub with the wine if you wish. Weigh the meat and record the result.

**5.** Find a corner that has skin and poke a hole through it, run a piece of butcher's string through it, and knot it. Hang the shoulder in the drying chamber for 4 to 6 months, or until it's lost about 30 percent of its weight.

**Yield: 1 spalla**

# Lardo

## (BACK FAT)

Lardo is fatback cured with salt and often herbs. As noted earlier, *lardo di Colonnata* has its own DOP—it's protected by law. Only lardo made in this town can use this name (though many break the law).

The special creaminess, the subtle flavors of *lardo di Colonnata* are the result of where the hogs are raised, what they eat, how they're handled by the several families of the town that make lardo, and, most notably, the marble they're cured in, the marble the town is literally built on. This linking of food and natural materials, commerce and daily life, is rare in America and is something we especially revere when we come across it. And so Brian and I try to bring our greatest care and respect to lardo when we prepare it here with our own hogs, with silent thanks to the generations reaching back a thousand years and more who lived and worked in Colonnata.

Lardo can be made from the back fat of any hog and taken in different seasoning directions. The fat from the Mangalitsas raised here makes amazing lardo; Durocs and Gloucester Old Spots can grow very fat as well. We offer two cures for lardo here.

Lardo can be served as is, as part of a salumi board. Or it can be served on a few leaves of basil and drizzled with fine extra virgin olive oil. It should always be thinly sliced—so thin that when draped over a warm toasted piece of bread, it melts ever so slightly, softening and heading toward translucency. It can be laid on pizza that's fresh from the oven (see page 229), or julienned and tossed with hot pasta and eggs and Parmigiano-Reggiano.

Lardo is about richness, and the creamy texture of fat and succulence, and all that is good.

Salume takes time to make and fresh fruit has to be perfectly ripened before you eat it. Give the meat too little time curing and it won't be delicious; leave the fruit on the vine too long and it will move past its prime. We love combining perfect fruit, well-cured meat, wine, and bread on a beautiful table, both simple and magical.

The big eight on display. Starting at the top: prosciutto, Calabrese salami (nice and fatty), coppa, lardo, spalla, guanciale, lonza, pancetta.

Making prosciutto can be a near religious experience. The ingredients are simple, the result complex and tantalizing. This ham hung in the transformation room for 450 days before Brian sliced it. A great ham starts with great pork, a good breed fed naturally, with nuts, fruit, and other food the pig foraged. The drying chamber contributes its own flavors, but it all starts with nature.

# Lardo

This is pure pig heaven, fat from the Mangalitsa pig seasoned with aromatic herbs and spices, then encased in salt for six months. Sliced paper thin, lardo is fantastic on a plate as is, on pizza or crostini, or whipped in a food processor into *spuma*. The knife belongs to Brian's mentor, Chef Milos Cihelka, a tool he used in his early days as a young chef.

Lonza, cured and air-dried pork loin, sliced to the point of transparency, lends itself to great eating and even better mouthfeel. Note the natural, good white mold on the outside, which should be brushed off before slicing.

# Pancetta

A staple in the Italian kitchen, pancetta can be salted and dried with skin on or off, rolled and tied, stuffed into a casing or left flat. Properly dry cured it can be sliced and eaten as is, like any salumi, or you can cut it into cubes and cook with it to add depth to countless dishes.

Speck, the boneless shoulder (spalla), salted, cold smoked over fruitwood, and air dried. Because it is made from heavily used muscle the flavor has excellent depth. Another way to use this cut is to salt and cure it with the bone in, like prosciutto.

Guanciale

Guanciale, or jowl, has some of the best tasting fat on the pig. The meat is delicious too, but there's more fat on the jowl than on pancetta. After salting, seasoning, and curing, guanciale becomes an invaluable flavor component in many dishes. Use it in any recipe that calls for pancetta. Or it can be pureed in a food processor into what's called *spuma di gota* (page 221).

In the summer of 1988, I left for a pleasure trip in Italy with my girlfriend of several months, a photographer. She was shooting a story on the famed marble of Carrara, on the central eastern coast of Italy. (Michelangelo's *David* was carved from its marble.)

We found a room in the town's sole hotel, Hotel Michelangelo, and discovered that the proprietor's son, Alex, spoke perfect English and would be happy to be our guide and interpreter. On one of several harrowing trips up those mountains to photograph the quarries, Alex took us to a friend's restaurant in the little village of Colonnata. Fausto Guadagni, the proprietor, was lanky, his dark hair unkempt and dark beard long as befitted someone with that name. He asked if he could serve us the local specialty—*lardo*. He said he cured it in his basement across the street. I asked what it was. Alex told me pork fat. *Crudo*.

"Raw?" I asked.

Alex nodded.

"You eat raw hog's fat?"

"*E bene*."

Remember, this was a time when Americans were only just beginning to come out of the culinary Dark Ages. Salami was made by Oscar Mayer and included by law a "kill stage," meaning it was cooked: pork was dangerous if it wasn't well-done. I'd never heard of prosciutto, let alone lardo. The verb *cure* had only medical meanings to us.

But when in Colonnata . . . , right? "Absolutely, we'll try that," we said.

Fausto brought out two plates each holding three paper-thin strips of the cured fat resting on a few basil leaves. Holding his thumb partially over the top of the olive oil bottle, he let fall a few streams and drops of gold onto the pearly white lardo.

We ate, staring at each other and smiling and eating more, and Alex and Fausto smiled too. It was one of those moments when I think, "This is not the world I woke up to." You *can* eat raw hog's fat. It's texture was sublime, melting as I chewed, dissolving on my tongue with the mild, pleasantly salty flavor of the hog and the olive oil, perfumed with the basil it rested on.

*continued on next page*

Donna was most surprised that it was smooth and creamy, more like butter than the fat surrounding a cooked pork chop. Fausto explained that it was crucial that the hogs were raised properly, that you had to know what the hog had been eating its whole life. Its diet, he said, was particular to this landscape. After lunch, he took us to his cellar and lifted the top off a marble cask, revealing huge slabs of fat curing in a salt-herb sludge.

I'm not a believer in a Big Man Upstairs pulling strings. I don't think that lardo, accepted like communion between Donna and me, foreordained a continuing collaboration, though it may have. I do know that it would have been really cool to time-travel that afternoon, just for a moment, twenty-two years into the future, when I would bring Brian to this mountain town, insisting that he had to taste the lardo.

It was 3:30 p.m., and Fausto's restaurant was closed when we arrived, as is all of Italy at that hour. I was able to peer in at the table where Donna and I had sat. A nearby salumière named Luigi, was, by grace, open, and we tasted his lardo, toured his cellar.

Brian stared, mouth agape, at the giant casks. Some, Luigi explained, were a hundred years old. He drew our attention to a smaller one in the corner, four hundred years old, he said.

"Holy F#$*, man," Brian whispered to me, "do you realize pork fat has been curing in that box since before the American Revolution?"

He marveled at the simplicity of it, fatback, salt, and herbs, and six months later, creamy, unbelievable lardo.

Soon we were zigzagging down the mountain, honking as we made each hairpin turn on the narrow road, and heading west toward Pescia to see our first Cinta Senese hogs foraging. It had been a transformative day for Brian. For me, with the memories of the girlfriend who became my wife, . . . well, every once in a while—it's very rare and probably a weakness—but once in a while I give in to the notion that it's at least *possible* there's a little more order to the universe than we can know.

This is why lardo is so important to me.

## Lardo in the Style of Colonnata

Colonnata is famous for the marble that's quarried there, and caskets made of the same marble serve as curing chambers for its lardo. The lardo is cured for a minimum of six months, packed in coarse sea salt loaded with the herbs that grow wild on the hillsides. It's possible that the calcium carbonate in the marble interacts with the salt, altering the quality of the brine and its effects on the fat.

Not surprisingly, we've been unable to source marble casks here in the United States, but if you can find one, or can bring one back from Colonnata, and have a cool place to store it, try this preparation and seasoning mixture! If you can't, you can still use the cure, following the instructions for Lardo Typico, page 101. We're giving specific quantities here, but if you have more or less back fat, the method is the same: pack the fat in salt with the aromatics and leave it alone in the dark for 6 months.

**THE CURE**

**About 2.5 pounds/1335 grams coarse sea salt or kosher salt**

**2 tablespoons/12 grams black peppercorns, toasted and roughly cracked in a mortar with a pestle or beneath a sauté pan**

**12 garlic cloves, minced**

**3 or 4 rosemary sprigs, crushed by hand**

**5 pounds/2270 grams thick fresh pork back fat, with skin, trimmed as necessary to fit into your marble cask**

**1.** Combine the cure ingredients, mixing well. Spread one-third of the mixture in a marble cask (or nonreactive container) that will hold the fat snugly. Add the fat, then add the remaining salt mixture, encasing the fat in salt.

**2.** Put the lid on the cask (or cover the container and wrap it in a black plastic bag to keep out the light) and store in a cool, dark place for 6 months.

*continued on next page*

**3.** Brush all the salt you can off the surface of the fat. Some will remain, giving the lardo a lovely little crunch. Slice off the skin and reserve, refrigerated, for another use (it will add body, and salt, to soups, stews, or stocks).

**Yield: 4 pounds/1800 grams lardo**

## Lardo Typico

Nothing compares to the lardo made in Colonnata, but if you cannot find a marble casket, this recipe is as close as you can get. Remember that the quality and the freshness of the fat are key. As is preventing light from reaching the fat during curing—light is fat's enemy. We give specific quantities here, but if you have more or less back fat, the method is the same: pack the fat in salt with the aromatics and leave it alone in the dark for 6 months.

**THE CURE**

**About 2.5 pounds/1335 grams coarse sea salt or kosher salt**

**2 tablespoons/12 grams black peppercorns, toasted and roughly cracked in a mortar with a pestle or beneath a sauté pan**

**12 garlic cloves, minced**

**3 or 4 rosemary sprigs, crushed by hand**

**2 tablespoons/18 grams juniper berries, crushed**

**1 tablespoon/3 grams finely ground dried bay leaves**

**5 pounds/2270 grams thick fresh pork back fat, with skin**

**1.** Combine the cure ingredients, mixing well. Spread one-third of the cure in the bottom of a nonreactive container (or a 2.5-gallon/9-liter zip-top plastic bag) that will hold the fat snugly. Pour the remaining cure over the fat, covering it completely.

**2.** Slide the container into a black plastic bag to prevent light from reaching the lardo and refrigerate for 6 months.

**3.** Brush all the salt you can off the surface of the fat. Some will remain, giving the lardo a lovely little crunch. Slice off the skin and reserve, refrigerated, for another use (it will add body, and salt, to soups, stews, and stocks).

**Yield: 4 pounds/1800 grams lardo**

# Lonza
## (LOIN)

Lonza is the boneless pork loin. It's the same cut of meat that's attached to the bones of pork chops. The muscle itself is completely lean, though you could ask whomever you're getting this from to leave plenty of back fat attached to the loin which will add flavor and succulence. Because it is often so lean, lonza (sometimes referred to as *lombo*) needs to be seasoned aggressively.

The tenderloin, which lies underneath the ribs toward the back of the animal, is a similarly lean cut, with very little fat on the exterior, that can be cured in the same way. It is called different names depending on where you are in Italy; here we refer to the tenderloin as filetto.

As always, the quality of the end result begins with the quality of the pork you purchase, and this is never truer than when using cuts as lean as the loin. Cure a loin from an American factory-raised hog, and you will get an all-but-flavorless lonza. Lonza from a well-raised hog fed on good grains, nuts, and mast will have a dense, chewy texture; a rich, deeply porky flavor; and a appealing dark red color.

Lonza and filetto are great cuts for curing at home because, like coppa, they're relatively small. (The lonza is, in fact, an extension of the coppa.) And leaner cuts are easier to cure at home than fattier ones, which present a greater danger of rancidity. You have to be careful about rancidity here; unlike the coppa, the fat of which lies mainly inside the muscle, the fat here is on the outside.

# Pepper-Cured Lonza

Because pepper is the only seasoning here, we urge you to toast and grind peppercorns in a spice mill or coffee grinder. If you want to get really serious about it, use a fine pepper such as Tellicherry, grown in India and left longer on the vine for greater flavor and size; Sarawak pepper from Borneo; or Javanese long pepper, which is especially spicy.

To toast peppercorns, put 2 to 3 tablespoons in a skillet over high heat. When the pepper begins to release its fragrance, after a couple of minutes or so, pour the peppercorns into a spice mill or coffee grinder and grind until they're pulverized. Pass the ground pepper through a medium strainer to catch any larger pieces. In the cure, we use a few tablespoons of pepper per pound/450 grams salt (it's OK to measure by sight). You'll need a tablespoon or two to coat a 5-pound/2270-gram loin.

**Boneless pork loin, heavy sinew removed (with some of the back fat left on if you wish)**

**THE CURE**

**Coarse sea salt or kosher salt**

**Black peppercorns, toasted and roughly cracked in a mortar with a pestle or beneath a sauté pan**

**Dry white wine for rinsing the meat (optional)**

**Black peppercorns, toasted and finely ground**

**1.** Combine the cure ingredients in a pan large enough to contain the loin, and roll the loin in the cure until it is uniformly coated (see The Salt-Box Method, pages 79–80). Put the loin in a 2.5-gallon/9-liter zip-top plastic bag. Add pepper, 1 or 2 tablespoons, or to taste. Squeeze as much air out of the bag as possible and seal the bag.

*continued on next page*

**2.** Put the loin on a baking sheet, put another pan on top, and weight it down with 8 pounds/3600 grams of weights. Refrigerate for 1 day per each 2 pounds/1000 grams. Midway through the curing time, flip the loin, redistributing the salt and pepper as you do so, and weight it down again.

**3.** Remove the loin from the bag, rinse under cold water, and pat dry. Rub with the wine if you wish. Weigh the pork loin and record the result.

**4.** Dust the loin with finely ground pepper, evenly coating all surfaces.

**5.** Tie the loin as you would a roast (see the illustrations on pages 122–25). If you're using weight to determine doneness, weigh the meat and record the result. Hang in the drying chamber for 3 to 4 weeks, or until it has lost 30 percent of its weight.

**Yield: 1 lonza**

The best technique for removing silverskin: angle the blade slightly upward while pulling the sinew away from the meat.

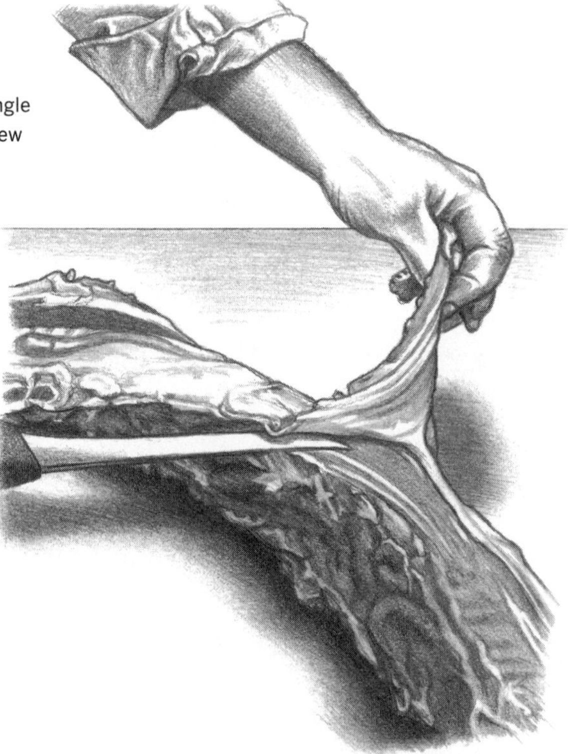

# Pancetta
## (BELLY)

If you were taking part in a debate and were forced to argue that there is a preparation in the salumi repertoire that is more valuable than pancetta, you'd have your work cut out for you. Certainly some salumi can deliver amazing pleasures to the palate, greater than the taste of the best pancetta. And both Spain and Italy produce dry-cured hams that are the most convincing argument for the existence of God I know of. But pancetta—cured hog belly—is unmatched in terms of flavor and versatility. With a sublime flavor when simply thinly sliced like prosciutto and salami, it can also be cooked and served, and it is every bit as satisfying as its American cousin, bacon, cured belly that's hot-smoked.

Braised or roasted, a slab of pancetta can be a main ingredient. Or it can be cut small and used as a base flavor. Ground or diced and heated to render its fat, it becomes the medium in which to sweat onions for a sauce, adding a depth of flavor that no other ingredient can. It can be dried flat (*tesa*) or rolled (*arrotolata*) or pressed between sticks (*steccata*), for different effects. In some areas in northeastern Italy, the belly is smoked, another variation on the versatile pancetta.

Finally, it's the most accessible salumi preparation for the home cook, something of a no-brainer that can become a constant presence in your kitchen.

Guanciale has uses similar to those of pancetta—it too can be fully cured, thinly sliced, and served. And it can be diced and sautéed and used as a base for sauces, or in a pasta sauce. But guanciale is more about the fat than the meat. Pancetta's ratio of roughly equal parts meat and fat makes it such a great ingredient and tool.

We recommend buying and curing skin-on belly. The skin is an amazing part of the hog—connective tissue, almost pure collagen, a protein that when broken down by cooking transforms into gelatin, which gives body to stocks and helps head-cheese set up. Cured skin takes on the same complex flavors as cured meat and fat, and it imparts those flavors to whatever you add it to. It can also be braised, then scraped of its fat and fried until crispy—known here as cracklings. In America, the skin is routinely removed from the hog because it's easier than leaving it on, scalding the hog, and removing the bristles with a scraper, so skin-on belly can be hard to find. Any type of pancetta can be made with skin-on or skin-off belly. The skin is very tough and needs long, moist cooking to break it down and tenderize it. Generally, rolled pancetta should be made with skinless belly, or when you roll it, some of that skin winds up in the center, making it hard to cook properly. However, the technique we use here calls for removing part of the skin, leaving just enough to wrap the pancetta.

Most belly comes in at around 10 pounds/4500 grams, so that's what these recipes call for, but they're easily halved. Because the weight will vary, we recommend salting it according to that weight at 3%. (In other words, multiply the weight in ounces or grams by 0.03 and add that amount of salt to the cure.)

Bellies are about 3 feet/1 meter long. If you're using a whole belly, you'll need a nonreactive container long enough to hold it lying flat. It can be halved so it fits in a 2.5-gallon/9-liter zip-top plastic bag. You can also fold it in half and cure it in the bag. If you do so, we recommend that you overhaul it every other day—that is, flip and rerub the belly to ensure that it gets thoroughly cured before being rolled and tied.

American bacon is made with sodium nitrite, which prevents the growth of botulism bacteria in the smokehouse and results in the pink color and piquant flavor we're familiar with. The *tesa* recipe here does not call for sodium nitrite, but we do recommend using this curing salt in the rolled pancetta, to prevent the possibility of the growth of botulism bacteria in the center of the roll.

The following are the main forms of pancetta you'll encounter in Italy. If you want to smoke it for the type of the pancetta made in the regions that were formerly part of Austria, use the cure for the rolled pancetta, which includes sodium nitrite.

# Pancetta Tesa

This is one of the easiest and most satisfying salumi preparations to make at home: belly cured with salt and lots of aromatics, rinsed, and hung to dry. That's all there is to it. You don't even need any special drying area: you can simply hang it from a hook in your kitchen for a couple weeks, and it will be fabulous (though if you have a drying chamber, by all means use it). And you don't have to dry it if you don't want to: maybe you'll be so hungry for it that you won't be able to wait and you'll want to cook and eat it right away. However, when you dry it, you intensify its flavor and make it a little more dense and chewy than if you were to cook it right away. Pancetta is usually used in cooking, but if you want to serve it with other salumi, thinly sliced and not cooked, we recommend following the basic dry-cure procedure and drying it until it has lost 30 percent of its original weight. Because we use a lot of aromatics in the cure, we're providing exact measurements, but feel free to use the salt-box method (see pages 79–80) to cure the belly if you wish.

Pork belly can come with skin or without. Buy it with the skin if possible. The skin will help the belly to keep its shape as it dries. But skin must be cooked in order to become tender; if slicing the belly to eat without cooking it, as for a salumi board, remove the skin. The skin can be added to stocks and soups for great body, or it can be braised until tender and fried for cracklings. If you're unable to buy skin-on belly, don't worry; the belly itself will be just as good.

*continued on next page*

**THE CURE**

4.8 ounces/135 grams sea salt (salt-box method, see pages 79–80, or 3% of the weight of the meat)

3 tablespoons/18 grams black peppercorns, toasted and roughly cracked in a mortar with a pestle or beneath a sauté pan

¼ cup/20 grams juniper berries, crushed

8 garlic cloves, minced

8 bay leaves, crumbled

8 rosemary sprigs

One 10-pound/4500-gram fresh pork belly

**1.** Combine the cure ingredients in a nonreactive container large enough to hold the belly flat (or use a zip-top plastic bag; see page 106). Add the belly and rub the cure all over it. Refrigerate for 5 days, flipping the meat and rerubbing it to redistribute the cure at least once, midway through curing.

**2.** Remove the belly from the pan and rinse off the cure ingredients under cold running water.

**3.** If you're using weight to determine doneness, weigh the meat and record the result. Poke a hole through one corner of the belly, run a piece of butcher's string through the hole, and knot it. Hang the belly for 2 to 3 weeks, or until it's lost 30 percent of its weight.

**Yield: One 7-pound/3000-gram pancetta**

# Pancetta Arrotolata

Rolled pancetta is what Americans think of when we think of pancetta. The thickness achieved by rolling it combined with a long drying time results in a deep, heady funkiness in the flavor. If you're unable to find skin-on belly, this can be made with skinless belly, though the skin provides an added measure of protection to the meat. We trim the skin so that there is only enough left to cover the exterior of the meat, with none wrapped inside the pancetta.

For rolled pancetta, the drying time is not critical. You can slice and cook pancetta immediately after it is cured. The hanging time (5–7 days), though, will deepen and enhance the flavor. If you want to be able to slice your pancetta very thin and serve it as is, you should dry it as you would any muscle, until it's lost about 30 percent of its weight. Because we use a lot of aromatics in the cure, we're providing exact measurements, but feel free to use the salt-box method (see pages 79–80) to cure the belly if you wish.

**THE CURE**

**4.8 ounces/135 grams sea salt (salt-box method, see pages 79–80, or 3% of the weight of the meat)**
**2 teaspoons/12 grams pink salt**
**3 tablespoons/18 grams black peppercorns, toasted and roughly cracked in a mortar with a pestle or beneath a sauté pan**
**¼ cup/50 grams packed brown sugar**
**¼ cup/20 grams juniper berries, crushed**
**8 garlic cloves, minced**
**8 bay leaves, crumbled**
**10 thyme sprigs**

**One 10-pound/4500-gram fresh pork belly with skin**
**½ cup/48 grams finely ground black pepper**

*continued on next page*

**1.** Combine the cure ingredients in a nonreactive container large enough to hold the belly flat (or use a zip-top plastic bag; see page 106). Add the belly and rub the cure all over it. Refrigerate for 5 days, flipping the meat and rerubbing it to redistribute the cure at least once, midway through curing.

**2.** Remove the belly from the pan and rinse off the cure ingredients under cold running water.

**3.** If you're using weight to determine doneness, weigh the meat and record the result. Dust the skin side with the pepper. Roll up the belly as tightly as possible and tie using a continuous tie (see the illustrations on pages 122–25).

**4.** Hang the belly in the drying chamber for 2 to 3 weeks.

**Yield: One 7-pound/3000-gram pancetta**

## Pancetta Steccata (Rolled and Pressed Between Sticks)

In Bologna, when you enter most salumerias, you will see, between the hanging prosciutto and culatello, whole pancettas, rolled and pressed between sticks. This technique facilitates the drying process and makes it oval. If your drying chamber is big enough to accommodate a full rolled belly, this is a fun method to try. You'll need two wooden dowels, each 1 inch (2.5 centimeters) thick and about a foot (30 centimeters) longer than your pancetta.

Follow the instructions for the pancetta *arrotolata* through the rolling and tying. Fasten battens on either side of the pancetta and tighten them with torniquets.

# Prosciutto
## (HAM/BACK LEG)

Prosciutto is the salted and air-dried back leg of the hog. The meat should be rosy in color, the flavor sweet, nutty, and hammy; the surrounding fat pearly and creamy. That's all there is to it, a big muscle, a little bit of fat. Slice it thin. Eat it.

And yet this dry-cured ham is the crowning glory of all hog preparations, of all that salumi offers, the ultimate expression of *terra*, of the hog, the power of salt, and the effects of drying meat over many months, often years. The experience of eating it, while intensely pleasurable, filled with unusual and beguiling aromas, also has another sense intermingling with the salty, sweet, nutty flavors: a sense of eating something forbidden. Because, really, this is a huge muscle that's never been cooked, that is, strictly speaking, raw down to the bone. We shouldn't be able to take so much pleasure in such a thing. And yet we do, because the pleasures of eating the finest dry-cured ham are unmatched.

The requirements to bring a ham to such a state depend on an extraordinary range of variables. The breed of the hog, how the hog is raised, where the hog grows, what it eats, how it's slaughtered, how it's handled after slaughter, how it's salted and for how long, where and how it dries, the quality of the air that carries its inner moisture away—all of these affect the final taste and fragrance of the meat. So diverse are Europe's great ham-producing areas, so nuanced are the flavors from varying regions, that these dry-cured hams (unlike pancetta or lonza or spalla or coppa) are designated by the region where the hogs grow, and the use of these regional names is often protected by law. *Prosciutto di Parma* from the heart of Italy's Emilia-Romagna—the salumi heart of the country—where the hogs are fed the whey that is a by-product of the region's famous Parmigiano-Reggiano, is distinct from the *Prosciutto di San Daniele*, from northeastern Italy, where the same breed of hog is fed a diet of corn, barley, spelt, wheat, and whey. And these are different from *jambon Bayonne*, from southwestern France, and the Spanish *jamón ibérico*, known by the breed, the black-footed

Iberian hog that roams vast pastures feasting on sweet wild acorns, dropped by the abundant oaks, and other mast native to the region.

Writing in the *New York Times* about the burgeoning American dry-cured ham scene, food-science authority Harold McGee outlined the requisites for achieving the deep porky flavors and aromas as wild and diverse as tropical fruit, caramel, and chocolate in dry-cured ham.

The hog, he writes, "should be mature, well fed, and free to run around. Muscles of such an animal are packed with the raw materials for creating flavor, enzymes that will catalyze the first stage of that creation, and fat to lend tenderness and moistness.

"Then there's time. It takes many months for muscle enzymes to break down flavorless proteins into savory amino acids, odorless fats into aromatic fragments, and for all these chemical bits and pieces to interact and generate new layers of flavor. And it takes months for meat to lose moisture and develop a density of flavor and texture."

Of special note is a small spot of earth, the environs of the city of Zibello along the Po River, where the finest form of prosciutto of all is cured. It's called culatello and is the king of kings. It's a big ball of meat cut out of the back of the ham, salt-cured then sewn up inside a hog's bladder, wrapped in a web of butcher's string and hung in a cellar for at least a year, or two years, some for nearly three years. This last is so close to rot that the pleasure of eating it feels thrillingly dangerous.

As with other large whole-muscle cuts, the curing rule is fairly standard: the ham is completely packed in salt and cured for one day for every kilogram. If only one ham is curing, it's often pressed down to help squeeze moisture toward the surface. Salumièri making multiple hams often stack one on top of the other to serve as the weight, rotating them throughout the cure. Single hams benefit from something as heavy as a cinderblock.

There's an important issue unique to preparing these hams. When the hog is slaughtered, it is bled out. Careful butchers do it in such a way that all the blood leaves the ham, but sometimes blood remains in the femoral artery, which runs along the inside of the femur. Blood that remains in a ham that is cured can become sour and affect the flavor of the ham. Always check to make sure there is no blood in the femoral

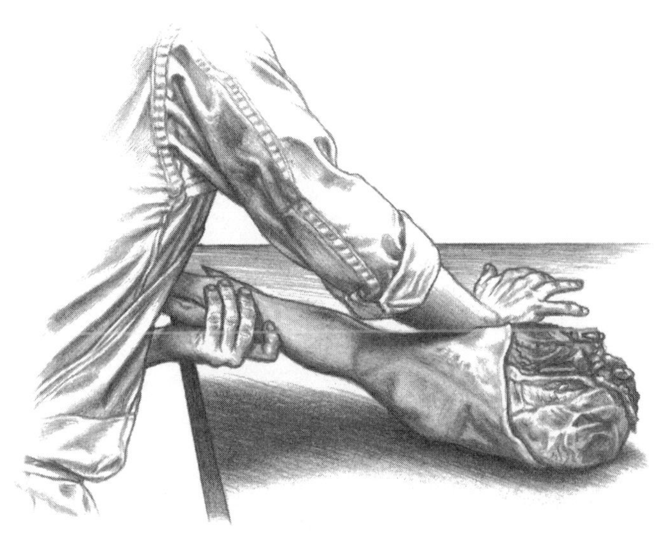

Depending on how your hog has been slaughtered, there may be some blood remaining in the femoral artery, which runs along the femur, the main leg bone of the ham. It's important to massage as much blood as possible out of the artery, as it will spoil if left in the ham. With the leg skin side down, push the heel of your hand from the shank up to the top of the ham, pressing out any remaining blood.

artery, and massage it out if there is, or if you're unsure whether or not there is blood in the artery. Press the heel of your hand all the way up the inner part of the leg (see the illustration to the left), until no more blood comes out.

Because of its size and the time it needs to dry properly, prosciutto is one of the most difficult cuts to cure successfully. It requires all those special qualities mentioned above, including the proper amount of time on the salt and good drying conditions, and it's a huge piece of meat to have hanging around for a year or more.

Can you go to your local supermarket, buy a fresh ham, and dry cure it? Sure. Will it be any good? Not likely.

There's a reason *jamón ibérico* and *prosciutto di Parma* are protected by the law. In the United States, you couldn't buy Iberian ham, until recently, but now, happily, you can, thanks to relaxed importing regulations. Even more happily, producers throughout the country are offering their own aged American hams (see Sources, pages 267–68), some of which experts judge the equal of the best of Spain, France, and Italy. And chefs offer it on their salumi boards, usually noting them on the menu by name: La Quercia, for example, which finishes some of its Berkshire hogs on acorns, like the *patas negras* of Spain, and Benton's Smoky Mountain Country Ham.

For all these reasons, prosciutto is one of the most satisfying cuts for the salumière to cure.

Hams can be cured on the bone as prosciutto, or they can be cured off the bone as culatello and fiocco (see pages 174–78). If you are determined to do a ham and your space is limited, one of the latter is the way to go.

## Prosciutto

This is the most demanding cut to cure because of its size, and for this reason is one of the most awesome and satisfying cuts for the cook who nails it. Try to source Berkshire, Duroc, or Gloucester Old Spots and a grower who is aware of your intention to dry cure the ham. The best farmers know how to handle their animals. Ask the farmer if he or she uses natural feed, whether the hogs graze, if there are acorns or fruit on the property. In the Midwest, hogs often finish where acorns and apples grow.

We can't overemphasize the importance of massaging the muscle to remove any blood remaining in the femoral artery (see pages 112–13). The blood will spoil in the ham if it's not removed.

Unless you're an expert butcher, we recommend that you leave the aitch bone attached to the ball joint. The aitch bone is part of the hog's pelvis, the bone to which the leg bone, the femur, is attached. It's traditionally removed, leaving the ball joint of the femur visible, but this also increases the odds that air pockets will be created, where molds can grow. You can remove the bone after the ham is dried.

The ham must be packed in salt and weighted down heavily so that as much moisture as possible is removed. We recommend using a large black plastic bag for the cure; it will keep the salt close to the ham. You'll need a sheet of wood or a baking sheet and weights that equal the weight of the ham.

In terms of drying, you'll need a chamber large enough to hang the ham. It shouldn't be laid on a flat surface or even a rack; it needs to hang. Also, the chamber should be dark most of the time, because light can make the fat rancid over the time it will take for the ham to cure.

To keep it from drying out, we spread a concoction called *strutto*, equal parts flour and lard, on the exposed flesh after about the first week of drying.

When the ham has finished curing and you're ready to taste it, remove the skin from the area you'll be carving, and slice the meat as thin as possible. It should be deeply rosy and have a dense, toothsome texture and a sweet, nutty, porky flavor. To store it, cover the cut surface with parchment paper or butcher's paper. While the ham is best the sooner it's eaten, it will keep in the refrigerator for months.

For a detailed description of tying the ham to hang it as well as carving it, see pages 122–25 and 44–46.

**1 bone-in hind leg of pork, skin on (about 24 pounds/11000 grams)**
**Several pounds of sea salt**
**About ½ cup/190 grams *strutto* (¼ cup/30 grams flour mixed with ¼ cup/ 60 grams lard)**

**1.** Pack the ham completely in salt in a large black plastic bag or large plastic bin. All the surfaces should be covered with salt.

**2.** Put a baking pan on the ham, weight it with about 24 pounds/11000 grams of weights, and refrigerate. Leave the ham in the salt for 1 day per each 2 pounds/1000 grams (in this case, 10 to 12 days). Inspect the ham every other day to make sure it's still completely packed in salt. If the ham has released so much liquid that it is sitting in a pool of it, pour off the liquid and add more salt.

**3.** Remove the ham from the salt, and rinse thoroughly under cold running water. Pour some dry white wine over it and rub off any remaining salt if you wish.

**4.** Weigh the ham and record its weight. Tie the ham at the hoof and hang it in the drying chamber. After a week or so, once the exposed flesh has dried somewhat, smear it with the *strutto*. Then leave it alone for a year, or until it has lost 30 percent of its original weight. It may be ready after 10 months, and a couple more months won't hurt it. (Depending on the conditions in your drying chamber, you may be able to leave it there for up to 2 years. The danger in letting it dry too long, though, is that it can be come dry, leathery, and tough.)

**Yield: One 17-pound/8000-gram prosciutto**

# Salami

If there is art in the wide-ranging craft of salumi, it is in the making of salami, that seemingly simple mixture of meat, fat, and aromatics, cured with salt, fermented, and dried. This is truly where the salumière makes his or her magic. The big hams (prosciutto and culatello), the other whole muscles (lonza, pancetta, and coppa), and the cured fat (guanciale and lardo) are almost wholly dependent on the hog itself. Sure, you have to salt whole muscles properly, and yes, you can alter their flavor by dusting the exterior or coating it with herbs, but it's in the making of a salame that the cook or chef must become fully engaged in order to transform trim, scraps, and bits of this and that into something that's eyes-closed-mmmmmmmmoan-inducing. The key here is the transformation.

It's not just the mixing together of ingredients and seasonings, like making a cake, or even making a regular sausage, though this is a fundamental stage of their creation. Unlike cakes and other sausages, salami are fermented. We know, of course, there are all kinds of fermented foods we love to eat, and these are all some of the most satisfying foods to make yourself: sourdough bread, sauerkraut and kimchee, and wine and beer. Cheeses rely on microorganisms and their salutary day jobs within our food. In India, the numerous bacteria that thrive in yogurt are believed to relieve the gastrointestinal distress caused by other nasty ones. There is some logic to this: like the good molds that prevent bad molds from taking over when curing meat, introducing good bacteria into a system may well reduce the prevalence of bad bacteria.

But fermented meat is a whole different situation, with both greater splendors and greater dangers. It *can* kill you if you don't pay attention and do it right. The botulism toxin is among the most deadly on earth. It was first identified and linked, as it were, to sausages in Germany in 1830, and named in 1875 after the Latin word for sausage, *botulus*. So take ground fermented meat that dries at near room temperature seriously. Happily, it's easy to make these creations safe and delicious. The botulism issue concerns only sausages—the anaerobic bacterium doesn't grow on the exterior of meat, it can only grow and produce the toxin inside the airless interior of a sausage (or a home-canned jar of beans, or garlic in olive oil, for instance).

You just need to follow a couple of rules to keep *Clostridium botulinum* from making you sick. (It's like scuba diving: if you follow the rules, it's always safe; it's only when you break a rule that you open yourself up to danger.) See page 76 for the key safety issues involved in dry curing.

Being well informed, and with our common sense always dialed to eleven, we are comfortable with the safety of our dry-cured meats and prefer to dwell on the splendor part. The splendors of fermented meats are protean. Yes, transformations happen in dry-cured whole muscles, due to salt, enzymes, and drying, but the transformations that turn a mash of salted ground trim into a solid, sliceable, translucent mosaic of meat and fat—that's the kind of work that gets a cook's blood flowing.

Salami are made from the trim left from the butchering of the animal, which is mixed with fat, ground, seasoned, stuffed into casings, and hung to dry. But to say "trim" is somewhat misleading. For really good dry-cured sausage, you don't want indiscriminate scraps: don't treat your sausage casing like a garbage can. You want *quality* trim, clean pieces of excellent meat and fat, all sinew removed. (If you watch your buddy taste your carefully made, beloved dry-cured sausage and then see him start picking something out of his teeth, that's the sinew you failed to remove.) As a rule, only put through your grinder what you would eat unground.

Starting with the simple premise of salami, countless variations can be made for countless sausages, depending on how the meat is ground (coarse or fine), how the fat is ground, and whether the meat and fat are ground together or separately. Is additional diced fat added? What kind of casings, fat or thin, are used? And, more obviously, what is the seasoning?—straightforward pork, garlicky, spicy, aromatic, sweet?

Pork is the meat most commonly used for dry curing because it tastes so good. There's no better meat for making salami, and no better fat. But other meats can and do make excellent salami. Wild boar, *cinghiale*, is a beloved meat for salami in Italy. Beef, venison, and lamb also work, but their fat does not—so for these sausages, it's always recommended that you use 75% lean meat and 25% pork fat.

Seasoning is critical. Salt, of course, is seasoning *numero uno*: it's critical for flavor and for safety. Proper salinity, between 2.3% and 3% (of the weight of the meat mixture) ensures proper curing and the prevention of bad microbes that can cause illness when allowed to grow inside warm ground meat. Black pepper is a common seasoning, as is garlic (though botulism spores, found in soil, can be carried on garlic). Wine can be added. These are what make a basic salame and, provided you have quality pork, salami doesn't need more done to it.

Not long ago, at a food gathering in Portland, Oregon, Michael attended a demonstration of how the French break down the hog (closer to the Italian style than the American). The butcher's dry-cured sausage was: meat, fat, 3% salt, and 0.2% ground black pepper. It was delicious (thank you, Kate Hill, for bringing the butcher, Dominique Chapolard, to America, and for smuggling the sausage past the American customs authority; Kate runs Camont, a cooking school-cum-butchery-program-cum-inn in Gascony). That simplicity is proof that it's the quality and care of the meat that is most important in the craft of making salami.

But, of course, you *can* do more to it than that. Add cayenne for a spicy salame. Add aromatics such as fennel seeds and orange zest. Add nuts or another internal garnish, such as diced cured meats.

There are two other ingredients that we feel are essential but may always be controversial: sodium nitrate and *Lactobacillus* bacteria. The nitrate's primary function is to prevent botulism contamination, and the bacteria generates acid to further protect the sausage as it cures. (These are covered in Dry Curing: The Basics, page 58.)

All of the salumi makers in Italy we spoke with (as well as Dominique Chapolard) denied adding either, though none could explain how they avoided the possibility of botulism poisoning. We've also spoken with Americans who have worked in Italy who say that while none of the salumièri will admit to using it, that they do nonetheless. Because we've been unable to find anyone who can explain how the possibility of botulism poisoning can be prevented absolutely without the use of a nitrate, and because all the food experts we've interviewed agree with us about the potential dangers of botulism, we don't recommend making sausage without it (whole muscles are different, again, because there is no way botulism-producing bacteria can get into the meat, so they don't need nitrate for safety reasons).

Moreover, the nitrate isn't bad for you. It converts to nitrite over time, which then converts to nitric oxide, so you are consuming little, if any, actual nitrite (and, again, nitrate, which becomes nitrite, is a valuable antimicrobial agent abundant in vegetables that works especially well in the acidic environment of our guts). As we've mentioned before, many commercial products use celery juice as a nitrite source, which allows food companies to market their product as "no added nitrites," but again, this is playing to a confused and misinformed consumer.

Bacteria is another story. All salami makers know that "good" bacteria feed on sugars and generate acid as a by-product, lowering the pH of the sausage (i.e., raising its acidity) to the point that the environment is inhospitable for bad bacteria—and also, happily, creating a tangy flavor we like. How these bacteria get into the sausage is where the contention lies. Most makers of quality artisanal salami in America use a commercially produced bacteria. It works well and consistently, but it can result in a salami that is too tangy for some traditionalists' tastes.

Salami makers in Europe rely on the bacteria already present on the surface of the meat and that, given specific saltiness, will thrive within the sausage. This practice results in a very fine flavor, but the disadvantage is that you can't control the bacteria, and so you may not create a sausage sufficiently acidic to cure properly. (See page 66 for more on using naturally occurring bacteria.)

We recommend you use a starter culture for both consistency and safety. For more on bacteria and bacterial starter cultures, see page 64.

## Our Basic Delicious, Simple Salame, the Starting Point for Countless Variations

The following recipe is the basis for other salami recipes in the book, but the technique is similar to that for any dry-cured sausage. It uses salt in the amount of 2.75% (0.0275) of the weight of the meat and fat, and it uses sodium nitrate (DQ Curing Salt #2) in the amount of 0.25% (0.0025) of the weight of the meat and fat. If you want to scale the recipe up or down, use the same percentages. We add additional fat to the shoulder butt in the amount of 15% to 25% of the weight of the shoulder butt.

The mixture can be stuffed into any size casing; the smaller the casing, the easier it will be to dry successfully. For most of the dry-cured sausages here, we use beef middles, which are about 2 inches/5 centimeters in diameter, cut into 18-inch/45-centimeter lengths. All are tied using a bubble knot, which prevents them from slipping out of the string while hanging. For more on dry-curing basics and creating an environment in which to dry cure, see pages 58–77. For the recommended equipment, see page 71.

**4 pounds/1815 grams fatty pork shoulder butt, cut into large dice, sinews and glands removed, and chilled until very cold**
**1 pound/450 grams pork back fat, cut into large dice**
**2 ounces/56 grams sea salt**
**1 teaspoon/7 grams DQ Curing Salt #2**
**2 teaspoons/4 grams black peppercorns, toasted and finely ground**
**¼ cup/60 milliliters chilled dry red wine**
**1 tablespoon/10 grams Bactoferm (live starter culture; see Sources, page 267)**
**2 tablespoons/30 milliliters distilled water**
**Mold 600 (see page 70 for more information and page 267 for sources) (optional)**

**Two 18-inch/45-centimeter lengths beef middle, soaked in tepid water for at least 20 minutes and rinsed**

**1.** Partially freeze the meat and fat.

**2.** Combine the shoulder, salt, curing salt, and black pepper and grind through a ⅜-inch/ 9-millimeter (large) die into the bowl of a stand mixer. Grind the fat into the bowl.

**3.** Using the paddle attachment, mix the ground meat and fat, adding the wine as you do so. Refrigerate the mixture in the bowl for 30 minutes.

**4.** Dissolve the Bactoferm in the distilled water. Using the attachment, mix the starter culture into the meat until well distributed.

**5.** Tie one end of each casing using a bubble knot (see page 122). Stuff the sausage into the casings, and tie each one off using another bubble knot. Using a clean needle, sausage pricker, or knife tip, poke holes all over the sausage, especially where there may be air pockets. If you're using weight to determine doneness, weigh the sausages and record the results.

**6.** Allow the sausages to incubate for 12 hours in a warm place (ideally at 80 degrees F./ 27 degrees C. and 80% humidity).

**7.** Hang the sausages in the drying chamber. If using the mold culture, mist them according to the package instructions. The salami are ready when they have lost about 30 percent of their raw weight.

**8.** Slice into a salame. It should be firm all the way through with an appealing deep red color and white dots of fat, all cohering in a lovely mosaic. If instead there's a ring of dark dried meat surrounding an interior that looks raw and mushy, you've had a case-hardening issue. After the sight evaluation, smell it: it should smell like salami. (If you've only encountered cooked factory-made salami from Oscar Mayer, buy some of the good stuff—see Sources, pages 267–68—and use it to compare.) If it looks good and smells good, peel off some of the casing, slice it thinly, and taste. If it tastes delicious, you've done it!

**Yield: Two 1.75-pound/800-gram salami**

## THE BUBBLE KNOT

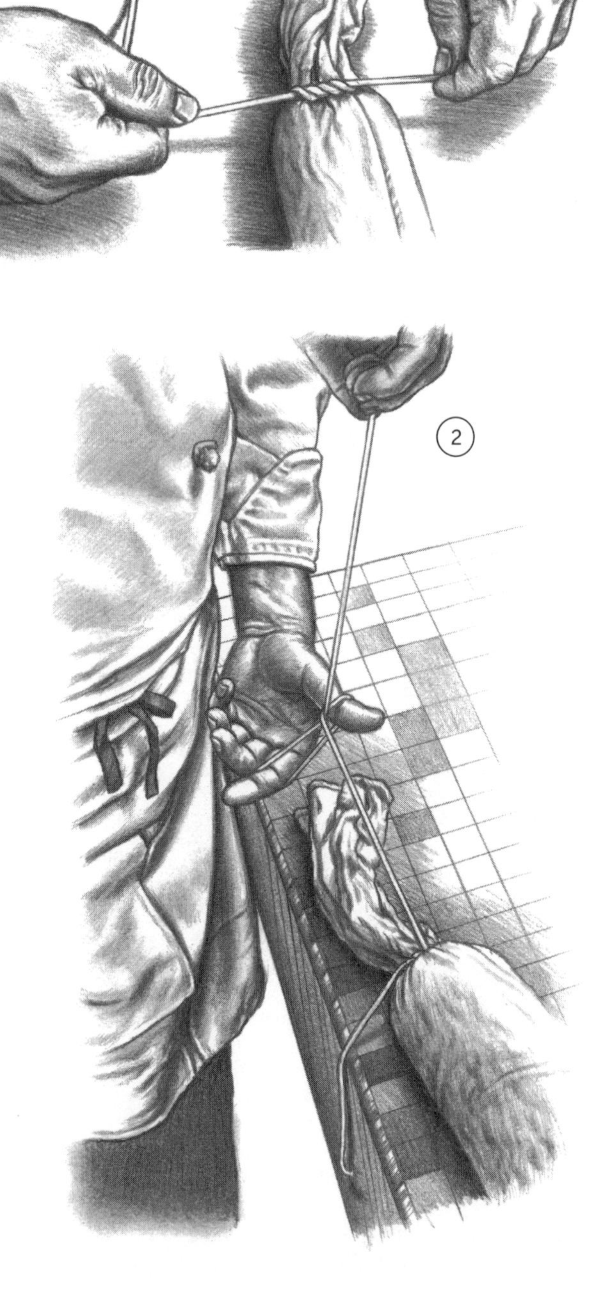


1. To tie the salame, measure off at least 18 inches/ 45 centimeters of string and cut the string from the spool. Make 3 loops of string, as if you were tying a square knot, only doing the first part of the square knot three times, and pull it tight. The triple looping will prevent the tie from loosening while you are finishing the knot. Tie the ends in a square knot. The right hand in this illustration should hold the long piece of string.

2. Loop the string around your thumb and finger and put the loop over the end of the casing, creating a hitch (the same as you would do for a continuous tie; see page 124).

3. Slide the hitch down toward your original square knot.

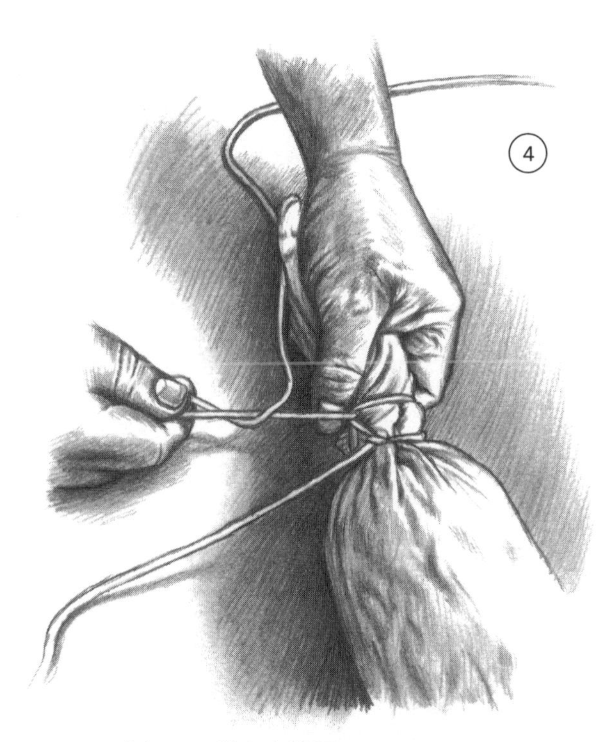


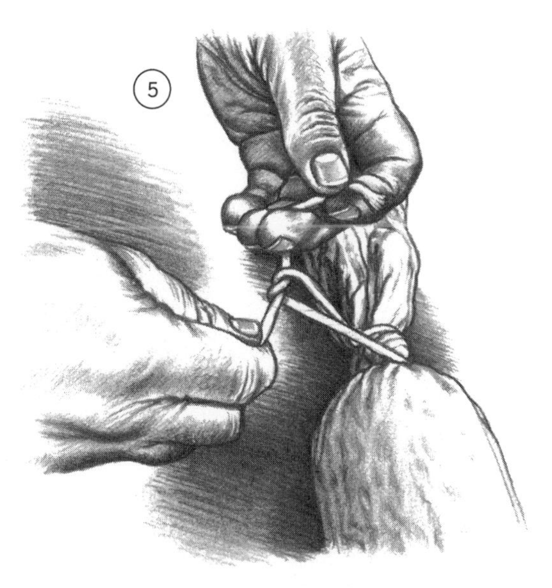


4. Leave a ½-inch/1.25-centimeter space between the knot and the hitch and pull the string tight, creating a bubble between the knot and the hitch.

5. To complete the knot, tie the two ends using a basic square knot.

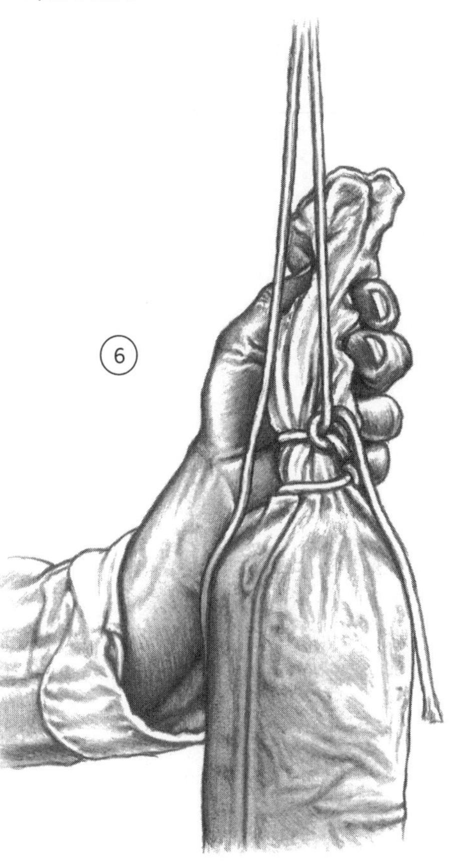


6. The finished bubble knot will be secure enough to hold the weight and the slippery casing so that your salame won't fall to your drying-chamber floor. Notice that there is still a gap between the 2 ties. Use the extra string to make a loop so that you can hang the sausage.

*continued on next page*

## THE CONTINUOUS TIE

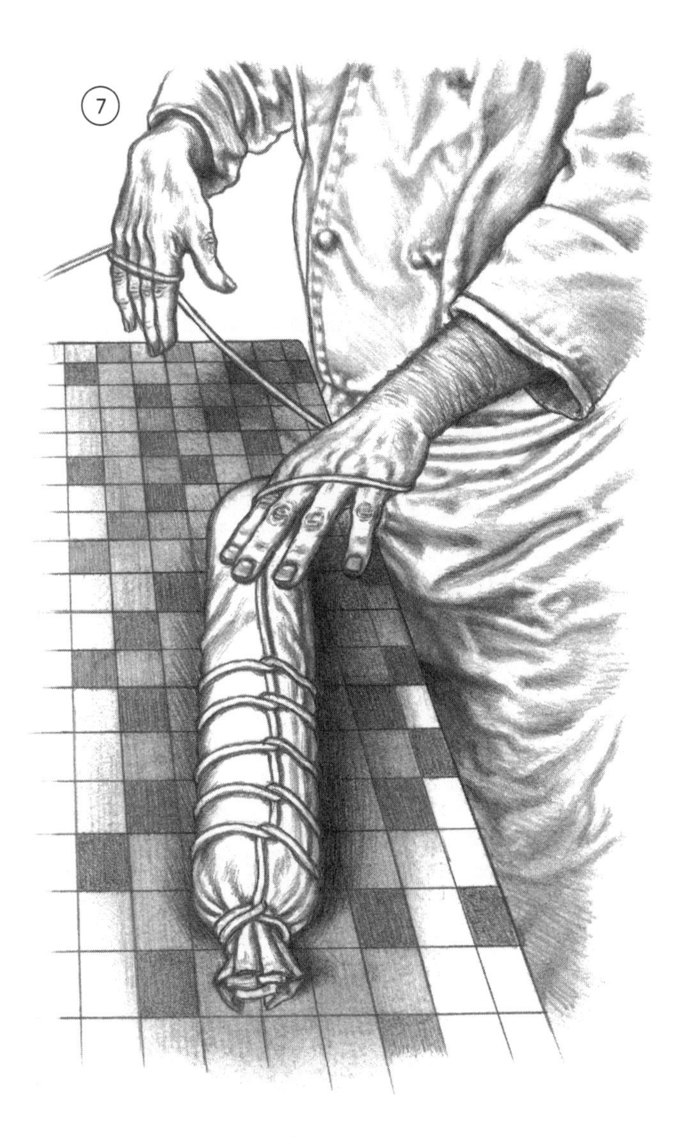


7. With another piece of string, tie a square knot at the base of the bubble knot, leaving the string still attached to the spool so that you have plenty of it. Then make a series of hitches: loop the string around your hand, twist your hand to make a loop, and slide the loop up to the top of the salame. Repeat this looping, bringing the next loop about 2 inches/5 centimeters below the first tie, then repeat the process down the length of the salame.

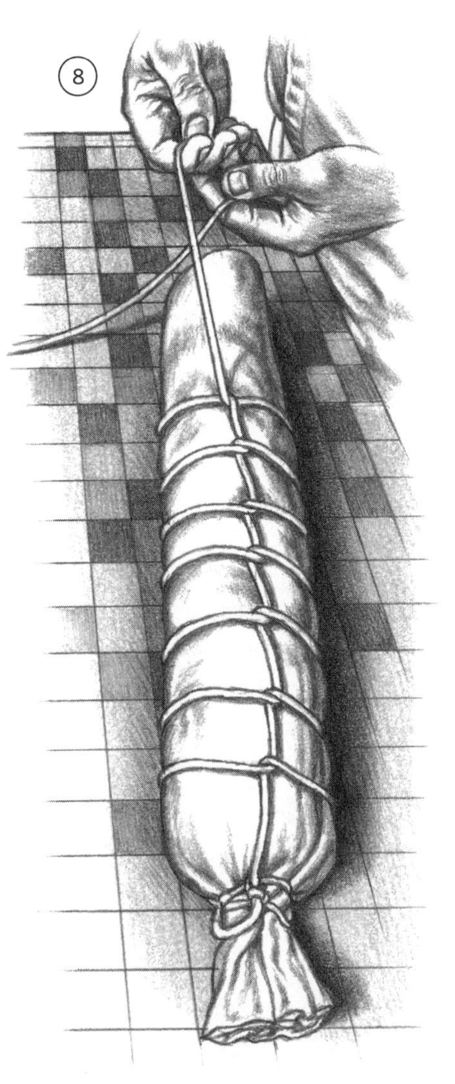


8. Pull the string tight to hold the salame securely. Be sure to position the string so that the loops are evenly spaced and the hitches line up.

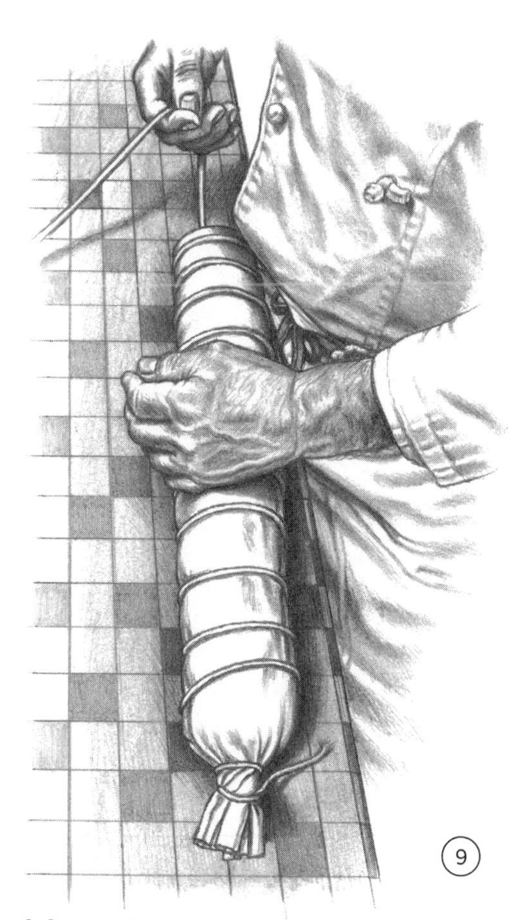


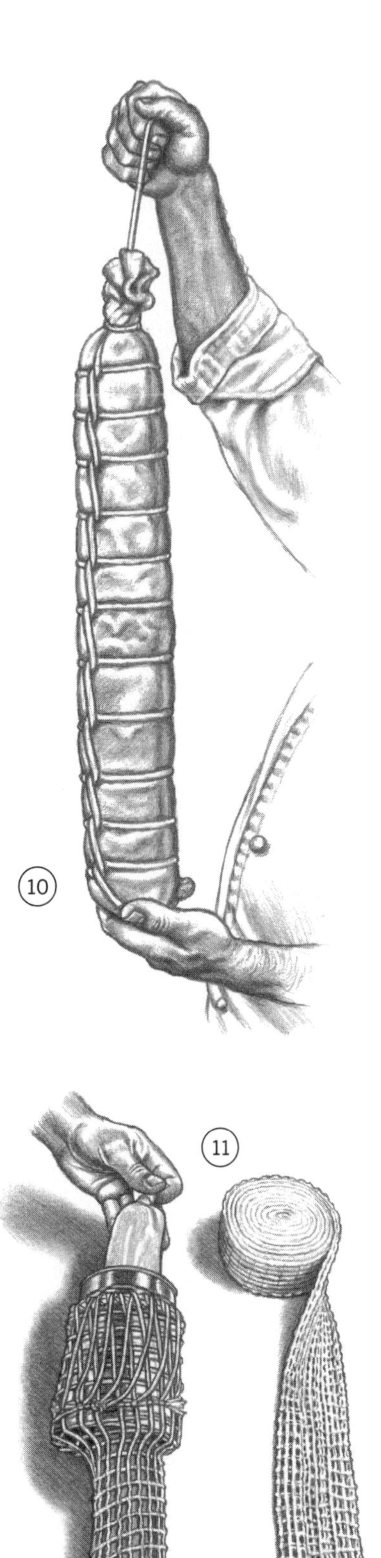


9. Once you've adjusted the spacing and pulled the string taut, turn the sausage over so that the hitches are on the bottom. Cut the string from the spool, leaving a piece that is the length of the sausage plus 18 inches/45 centimeters. Loop the string under and around each tie all the way up this side so that the sausage is secure on both sides. After looping the string around the final tie, wrap the end of the string around the bubble knot, loop it around itself, and pull it tight, do this again to cinch it, and tie it off.

10. You should have about 18 inches/45 centimeters of string free to make a loop with which to hang the sausage.

11. Netting is an alternative to the continuous-tie technique (see Sources, page 267). It comes in various sizes, so choose the right size for the item you're hanging. Slide the netting over an open-ended cylinder, leaving some netting hanging over one end, and slide the sausage through the cylinder, allowing the netting to encase it.

# 4. Deeper into the Craft of Dry Curing and Preserving Meat

In the first part of this book, we simplified salumi by breaking it down into its eight basic components. In this section, we offer variations on those eight themes. Variations are literally infinite and should always be to your tastes.

We've broken this section down into three parts: ground and dry cured (salami), whole and dry cured (e.g., coppa and prosciutto), and cooked (both ground and whole).

## Salami

The recipes in this section all involve ground meat. If you're just starting out, we recommend beginning with the basic salami recipe on page 120; it's exquisite and provides an excellent baseline from which to judge all your own dry-cured products. As a general rule, we encourage simplicity in salami, as in all culinary matters: salt and pepper and quality pork are really all you need. But also, as cooks, we want diversity, creativity, and variety. And so we offer the following variations on the basic dry-cured salami. We love these recipes and encourage you to try them and to experiment with your own variations. But it's important that you look at them as variations on a theme rather than isolated recipes.

The variations are many: some have interior garnishes—pistachios or whole peppercorns for instance—some contain elaborate spice mixtures, some use venison (here in America, we have big population of deer hunters eager to find new ways to use the abundance they find themselves with each fall). And there's a special type of spreadable salami called *nduja*, native to southern Italy.

Please read the salami chapter, pages 116–25, for important safety and technical information on making any salami.

# Finocchiona

Finocchiona (*finocchio* is Italian for fennel) is a sausage aggressively seasoned with fennel seeds, toasted and cracked. Make sure they're good ones, not seeds that have been sitting in your spice rack for who knows how long. This sausage was invented by a thief, a scoundrel who stole salami from a shop in Chianti. Pursued by the salumière, the thief hid the sausage in a patch of wild fennel. When he returned to retrieve his booty, he found that the fennel had given it a lovely and complex flavor, and thus was finocchiona born. Or so the story goes. The truth may be more mundane: peppercorns, which were imported, were more expensive in earlier times than the fennel that grew wild all over the place. And fennel gives a strong floral appeal to dry-cured pork. Although finocchiona is made all over Italy, the Chianti region is especially well known for it. Seek out (or forage) wild fennel seeds if you can; they have a more pungent and distinctive flavor than grocery-store fennel seeds.

**3.5 pounds/1590 grams lean pork shoulder butt, cut into large dice**
**1.5 pounds/680 grams pork back fat, cut into large dice**
**2 ounces/56 grams sea salt**
**1 teaspoon/7 grams DQ Curing Salt #2**
**1 tablespoon/6 grams coarsely ground black pepper**
**1 tablespoon/10 grams dextrose**
**2 tablespoons/12 grams fennel seeds, toasted and cracked**
**4 garlic cloves, minced**
**½ cup/125 milliliters chilled dry red wine, such as Chianti**
**1 tablespoon/10 grams Bactoferm (live starter culture; see Sources, page 267)**
**2 tablespoons/30 milliliters distilled water**
**Mold 600 (see page 70 for more information and page 267 for sources) (optional)**

**Two 18-inch/45-centimeter lengths beef middle, soaked in tepid water for at least 20 minutes and rinsed**

*continued on next page*

**1.** Partially freeze the meat and fat.

**2.** Grind the pork through a ½-inch/12-millimeter (extra-large) die into the bowl of a stand mixer. Grind the fat into the bowl. Add the salt, curing salt, pepper, dextrose, fennel seeds, and garlic and, using the paddle attachment, mix on medium speed, adding the wine as you do so, until the ingredients are well distributed and the meat is tacky.

**3.** Dissolve the Bactoferm in the distilled water, add it to the meat, and mix until well distributed, another minute or so.

**4.** Tie one end of each casing using a bubble knot (see page 122). Stuff the sausage into the casings, and tie each one off using another bubble knot. Using a clean needle, sausage pricker, or knife tip, poke holes all over the sausage, especially where there may be air pockets. If you're using weight to determine doneness, weigh the sausages and record the results.

**5.** Allow the sausages to incubate for 12 hours in a warm place (ideally 80 degrees F./27 degrees C. and 80% humidity).

**6.** Hang the sausages in the drying chamber. If using the mold culture, mist them according to the package instructions. The salami are ready when they have lost about 30 percent of their raw weight, about 3 weeks.

**Yield: Two 1.75-pound/800-gram salami**

## Salami Picante

*Salami picante*, spicy salami, is made with hot red pepper flakes. It's common in southern Italy. It's not pepperoni, which is as American as our "Italian" sausage. (Both were likely first made by immigrants and imitated by others in this country.)

**3.5 pounds/1580 grams lean pork shoulder butt, cut into large dice**
**1.5 pounds/680 grams pork back fat, cut into large dice**
**2 ounces/56 grams sea salt**
**1 teaspoon/7 grams DQ Curing Salt #2**
**2 teaspoons/6 grams freshly ground black pepper**
**1 tablespoon/7 grams red pepper flakes**
**1 tablespoon/8 grams fennel seeds**
**1 tablespoon/10 grams dextrose**
**3 garlic cloves, minced**
**½ cup/125 milliliters chilled dry red wine**
**1 tablespoon/10 grams Bactoferm (live starter culture; see Sources, page 267)**
**2 tablespoons/30 milliliters distilled water**
**Mold 600 (see page 70 for more information and page 267 for sources) (optional)**

**Two 18-inch/45-centimeter lengths beef middle, soaked in tepid water for at least 20 minutes and rinsed**

1. Partially freeze the meat and fat.

2. Grind the pork through a ¼-inch/6-millimeter (medium) die into the bowl of a stand mixer. Grind the fat into the bowl. Add the salt, curing salt, black pepper, red pepper flakes, fennel seeds, dextrose, and garlic and mix with the paddle attachment on medium speed, adding the wine as you do so, until the ingredients are well distributed and the meat is tacky.

*continued on next page*

**3.** Dissolve the Bactoferm in the distilled water, add it to the sausage, and mix for another minute, until well distributed.

**4.** Tie one end of each casing using a bubble knot (see page 122). Stuff the sausage into the casings and tie each one off with another bubble knot. If you're using weight to determine doneness, weigh the sausages and record the results.

**5.** Allow the sausages to incubate for 12 hours in a warm place (ideally 80 degrees F./ 27 degrees C. and 80% humidity).

**6.** Hang the sausages in the drying chamber. If using the mold culture, mist them according to the package instructions. The salami are ready when they have lost about 30 percent of their raw weight, about 3 weeks.

**Yield: Two 1.75-pound/800-gram salami**

## Orange and Walnut Salami

Brian created this sausage after a student asked him about putting walnuts rather than the customary pistachios in a pâté. Pistachios are great in pâté, but Brian thought that walnuts wouldn't work. The idea stuck in his head, though—what kind of *farce*, or meat stuffing, would they work in? They're a good, flavorful nut—maybe in a dry-cured sausage? And they are a fall nut, hog time, so it made seasonal sense.

Then it was a matter of pairing this flavor with complementary ones. He thought orange and walnuts, then he thought orange and anise. He tasted the sausage, and it reminded him of Christmas, with a refreshing sweet, spicy, fruity flavor and aroma. The orange zest makes a big impact, and the walnuts lend depth and textural crunch to the sausage. Great around the holidays, this is kind of like the pork version of fruitcake. In fact, if you're part of a family that feels fruitcake is obligatory during the holidays, we suggest you make and offer this instead!

4 pounds/1815 grams lean pork shoulder butt, cut into large dice
1 pound/450 grams pork back fat, cut into large dice
2 ounces/56 grams sea salt
1 teaspoon/7 grams DQ Curing Salt #2
3 tablespoons/18 grams sweet paprika
1 tablespoon/6 grams fennel pollen (see Sources, page 268)
1 tablespoon/7 grams anise seeds
1 tablespoon/6 grams cayenne pepper
1 tablespoon/12 grams dextrose
3 garlic cloves, minced
Grated zest of 1 orange
1 tablespoon/10 grams Bactoferm (live starter culture; see Sources, page 267)
2 tablespoons/30 milliliters distilled water
1 cup/120 grams walnuts, coarsely chopped
Mold 600 (see page 70 for more information and page 267 for sources) (optional)

Two 18-inch/45-centimeter lengths beef middle, soaked in tepid water for at least 20 minutes and rinsed

*continued on next page*

**1.** Partially freeze the meat and fat.

**2.** Combine the salt, curing salt, paprika, fennel pollen, anise seeds, cayenne, dextrose, and garlic and sprinkle over the chilled meat and fat. Add the orange zest. Grind the meat and fat through a ⅜-inch/9-millimeter (large) die into the bowl of a stand mixer.

**3.** Dissolve the Bactoferm in the distilled water. Using the paddle attachment, mix the pork on medium speed for 1 minute. Add the dissolved starter culture and walnuts and mix until the ingredients are well distributed and the meat is tacky.

**4.** Tie one end of each casing using a bubble knot (see page 122). Stuff the sausage into the casings, and tie each one off using another bubble knot. Using a clean needle, sausage pricker, or knife tip, poke holes all over the sausage, especially where there may be air pockets. If you're using weight to determine doneness, weigh the sausages and record the results.

**5.** Allow the sausages to incubate for 12 hours in a warm place (ideally 80 degrees F./ 27 degrees C. and 80% humidity).

**6.** Hang the sausages in the drying chamber. If using the mold culture, mist them according to the package instructions. The salami are ready when they have lost about 30 percent of their raw weight, about 3 weeks.

**Yield: Two 1.75-pound/800-gram salami**

## Porcini Salami

A good chef is by nature a penny-pinching miser, and Brian is not just a good chef, he's a great one, so dial that up several notches! Seriously, food costs are always on a chef's mind. Throwing away anything that might be used drives chefs crazy. Brian loves shiitake mushrooms and uses a lot of them in his cooking, but he hated throwing away pounds of stems. The stems are tough, chewy, and woody—there's no pleasure in eating them. But that didn't keep Brian from wanting to figure out a way to use them!

He let them dry naturally until they were completely dessicated and then ground them to a powder in a spice mill. Voilà: a fabulously versatile mushroom dust to flavor bread, add to pasta dough, or use in risotto. Brian learned to love all mushroom powders, which is how the great European mushroom, porcini, worked its way into this sausage. Most of us don't have ready access to the expensive, meaty fresh porcini—they are often sold dried and work well that way. We love the aggressive, earthy flavor of porcini, but you can substitute other dried mushrooms if you wish.

4 pounds/1815 grams pork shoulder butt, cut into large dice
1 pound/450 grams pork back fat, cut into large dice
2 ounces/56 grams sea salt
1 teaspoon/7 grams DQ Curing Salt #2
1 tablespoon/8 grams finely ground black pepper
1 tablespoon/6 grams ground dried sage
½ ounce/15 grams dried porcini mushrooms, finely ground (about 3 tablespoons)
½ ounce/12 grams garlic, minced (about 4, depending on size)
4 ounces/170 grams freshly grated Parmigiano-Reggiano cheese (about 1 cup)
¾ cup/175 milliliters chilled dry white wine
1 tablespoon/10 grams Bactoferm (live starter culture; see Sources, page 267)
2 tablespoons/30 milliliters distilled water
Mold 600 (see page 70 for more information and page 267 for sources) (optional)

Two 18-inch/45-centimeter lengths beef middle, soaked in tepid water for at least 20 minutes and rinsed

*continued on next page*

1. Partially freeze pork and fat.

2. Combine all pork, fat, salt, curing salt, pepper, sage, porcini, and garlic and grind through a ¼-inch/6-millimeter (medium) die into the bowl of a stand mixer.

3. Using the paddle attachment, mix the meat for a minute on medium speed, adding the wine and cheese as you do so, until the ingredients are well distributed.

4. Dissolve the Bactoferm in the distilled water, add it to the meat, and mix for another minute, until well distributed and the meat is tacky.

5. Tie one end of each casing using a bubble knot (see page 122). Stuff the sausage into the casings, and tie each one off using another bubble knot. Using a clean needle, sausage pricker, or knife tip, poke holes all over the sausage, especially where there may be air pockets. If you're using weight to determine doneness, weigh the sausages and record the results.

6. Allow the sausages to incubate for 12 hours in a warm place (ideally 80 degrees F./ 27 degrees C. and 80% humidity).

7. Hang the sausages in the drying chamber. If using the mold culture, mist them according to the package instructions. The salami are ready when they have lost about 30 percent of their raw weight, about 3 weeks.

**Yield: Two 1.75-pound/800-gram salami**

## Salami Tartuffi

Adding truffles to any food product is a good thing. Fresh white truffle peelings could be used here, but they are rather costly. Truffle oil is the next best thing—when used judiciously. It's strong, and you can go overboard with it. And a decade ago, every chef and his grandmother seemed to be pouring truffle oil onto everything in sight. But if you use a good brand, such as Urbani, and use it delicately, it is still a good product. Here it gives the salami a lovely aroma and flavor.

**4 pounds/1815 grams pork shoulder butt, cut into large dice**
**1 pound/450 grams pork back fat, cut into large dice**
**2 ounces/56 grams sea salt**
**1 teaspoon/7 grams DQ Curing Salt #2**
**1 tablespoon/8 grams finely ground black pepper**
**¼ cup/56 grams white truffle oil**
**½ ounce/15 grams garlic, minced (4 cloves)**
**¾ cup/175 milliliters chilled dry white wine**
**1 tablespoon/10 grams Bactoferm (live starter culture; see Sources, page 267)**
**2 tablespoons/30 milliliters distilled water**
**Mold 600 (see page 70 for more information and page 267 for sources) (optional)**

**Two 18-inch/45-centimeter lengths beef middle, soaked in tepid water for at least 20 minutes and rinsed**

1. Partially freeze the pork and fat, separately.

2. Combine the pork, fat, salt, curing salt, pepper, truffle oil, and garlic and grind through a ¼-inch/6-millimeter (medium) die into the bowl of a stand mixer using the paddle attachment, mix on medium speed, adding the wine as you do so, until the ingredients are evenly distributed.

*continued on next page*

3. Dissolve the Bactoferm in the distilled water, and add it to the meat, and mix for 1 minute, until well distributed and the meat is tacky.

4. Tie one end of each casing using a bubble knot (see page 122). Stuff the sausage into the casings, and tie each one off using another bubble knot. Using a clean needle, sausage pricker, or knife tip, poke holes all over the sausage, especially where there may be air pockets. If you're using weight to determine doneness, weigh the sausages and record the result. Allow the sausages to incubate for 12 hours in a warm place (ideally 80 degrees F./27 degrees C. and 80% humidity).

5. Hang the sausages in the drying chamber. If using the mold culture, mist them according to the package instructions. The salami are ready when they have lost about 30 percent of their raw weight, about 3 weeks.

**Yield: Two 1.75-pound/800-gram salami**

## Salamini Cacciatore (Hunter's Sausage)

Lightweight, easy to carry, and high in protein and flavor, these sausages are popular with hunters, hence the name. Very dense and chewy after they are dried, they hang in salumerias all over Italy, big long bunches of ropes of them, sometimes from the ceiling nearly to the floor. Usually thin butcher's string is used to tie off each sausage, and then the string is wrapped over the links to lend support so the upper casings won't break under the weight of fifty or sixty links. Tying them this way is a real craft.

This recipe calls for pork, pork fat, and boar; if you're a hunter and have done well, use venison. If you don't have access to boar or venison, you can use beef instead for this highly spiced sausage.

**2 pounds/900 grams pork shoulder butt, cut into large dice**
**2 pounds/900 grams wild boar or venison, trimmed of fat, or boneless beef chuck, cut into large dice**
**2 ounces/56 grams sea salt**
**1 teaspoon/7 grams DQ Curing Salt #2**
**1 tablespoon/6 grams coarsely ground black pepper**
**1 tablespoon/9 grams coriander seeds, toasted and coarsely ground**
**2 teaspoons/4 grams cayenne pepper**
**2 teaspoons/3 grams ground mace**
**1 tablespoon/10 grams dextrose**
**1 pound/450 grams pork back fat, cut into large dice**
**½ cup/120 milliliters dry white wine**
**1 tablespoon/10 grams Bactoferm (live starter culture; see Sources, page 267)**
**2 tablespoons/30 milliliters distilled water**

**10 feet/3 meters hog casing, soaked in tepid water for at least 20 minutes and rinsed**

*continued on next page*

**1.** Partially freeze the meat and fat.

**2.** Combine the pork, beef, salt, curing salt, pepper, coriander, cayenne pepper, mace, and dextrose and grind through a ½-inch/12-millimeter (extra-large) die into a bowl. Cover with plastic wrap and refrigerate it for at least 8 hours, and up to 24 hours.

**3.** Add the fat to the ground meat mixture and grind it all through a ¼-inch/6-millimeter (medium) die into the bowl of a stand mixer. Using the paddle attachment, mix on medium speed, adding the wine as you do so, until the ingredients are well distributed.

**4.** Dissolve the Bactoferm in the distilled water, add it to the meat and mix until well distributed and the meat is tacky.

**5.** Stuff the sausage into the hog casings and twist off into 5-inch/12-centimeter links. Using a clean needle, sausage pricker, or clean knife tip, poke holes all over, especially where there may be air pockets. If you're using weight to determine doneness, weigh the sausages and record the results.

**6.** Hang the sausages to incubate for about 24 hours in a warm place (ideally 80 degrees F./29 degrees C. and 80% humidity).

**7.** Hang the sausages in the drying chamber. The salami are ready when they have lost about 30 percent of their raw weight, about 3 weeks.

**Yield: 3.5 pounds/1500 grams salamini; about twenty 5-inch/ 12-centimeter sausages**

## Salami Ungherese

Brian's mom-in-law is Hungarian, and she often brought him sausages made by her Hungarian butcher. He could never figure out why he couldn't get his as flavorful—until she brought him a gift of fresh Hungarian paprika. The flavor was amazing, and it was powerfully clear to him how critical the source and quality of your dried spices are.

The name of this sausage might indicate that it is from Hungary, but it is actually produced in the northern part of Italy. It is a historical reminder that Austria occupied a part of northern Italy well into the twentieth century. The Austrians, and the neighboring Hungarians, were big into paprika and smoke. The technique for this smoked and dried sausage is similar to that for the Milanese style of salami, in which all the meat and fat are ground on the finest die, allowing the fat to be interspersed throughout and giving the salami a speckled appearance. It is cold-smoked and then hung to dry (see page 75 for information and issues concerning cold-smoking). Use the best Hungarian paprika you can find (see Sources, page 268).

**3.5 pounds/1590 grams pork shoulder butt, cut into large dice**
**0.5 pound/225 grams lean beef chuck, cut into large dice**
**1 pound/450 grams pork back fat, cut into large dice**
**7 tablespoons/42 grams sweet Hungarian paprika**
**2 ounces/56 grams sea salt**
**1 teaspoon/7 grams DQ Curing Salt #2**
**2 tablespoons/12 grams freshly ground black pepper**
**3 garlic cloves, minced**
**1 tablespoon/12 grams dextrose**
**¼ cup/60 milliliters chilled Tokay (Tokaji) or other sweet dessert wine**
**1 tablespoon/10 grams Bactoferm (live starter culture; see Sources, page 267)**
**2 tablespoons/30 milliliters distilled water**

**Two 18-inch/45-centimeter lengths beef middle, soaked in tepid water for at least 20 minutes and rinsed**

*continued on next page*

**1.** Partially freeze the meat and fat.

**2.** Combine the pork, beef, fat, paprika, salt, curing salt, pepper, garlic, and dextrose and grind them through a ¼-inch/6-millimeter (medium) die into the bowl of a stand mixer. Cover with plastic wrap and refrigerate until thoroughly chilled, 4 to 6 hours.

**3.** Using the paddle attachment, mix the ground meat on medium speed for 1 minute adding the wine as you do so.

**4.** Dissolve the Bactoferm in the distilled water, add to the meat, and mix until it's well distributed, and the meat is tacky.

**5.** Tie one end of each casing using a bubble knot (see page 122). Stuff the sausage into the casings, and tie each one off using another bubble knot. Using a clean needle, sausage pricker, or knife tip, poke holes all over the sausage, especially where there may be air pockets. If you're using weight to determine doneness, weigh the sausages and record the results.

**6.** Allow the sausages to incubate for 12 hours in a warm place (ideally 80 degrees F./27 degrees C. and 80% humidity).

**7.** Cold-smoke the salami for 5 hours at 80 to 90 degrees F./27 to 32 degrees C. (see page 75).

**8.** Hang the sausages in the drying chamber. The salami are ready when they have lost about 30 percent of their raw weight, about 3 weeks.

**Yield: Two 1.75-pound/800-gram salami**

# Soppressata Roman-Style

In Italy, soppressata varies from region to region. *Pressato* is Italian for pressed, and a fat salame can be pressed into an oval shape to facilitate drying, as we do here. Tuscan soppressata is not salami at all but rather salume made from meat, fat, skin from the head, and other trim, cooked with aromatics in a liquid until everything is tender. The mixture is stuffed into a large casing, and the gelatinous stock holding all the morsels together in a large casing will be sliceable when chilled; see page 184 for a recipe. Most often, though, soppressata is a dry-cured salame, as it is here, spicy from pepper and chile flakes, and with the fat ground through a larger die than the meat so the chunks are big and distinct.

**4 pounds/1815 grams pork shoulder butt, cut into large dice**
**1 pound/450 grams pork back fat, cut into large dice**
**2 ounces/56 grams sea salt**
**1 teaspoon/7 grams DQ Curing Salt #2**
**1 tablespoon/7 grams finely ground black pepper**
**1 tablespoon/7 grams red pepper flakes**
**1 tablespoon/9 grams dextrose**
**½ cup/125 milliliters dry white wine**
**1 tablespoon/10 grams Bactoferm (live starter culture; see Sources, page 267)**
**2 tablespoons/30 milliliters distilled water**
**Mold 600 (see page 70 for more information and page 267 for sources) (optional)**

**Two 18-inch/45-centimeter lengths beef middle, soaked in tepid water for at least 20 minutes and rinsed**

**1.** Partially freeze the meat and fat.

**2.** Grind the pork through a ¼-inch/6-millimeter (medium) die into the bowl of a stand mixer. Grind the fat through a ⅜-inch/9-millimeter (large) die into the bowl.

*continued on next page*

**3.** Add the salt, curing salt, pepper, red pepper flakes, and dextrose and mix with the paddle attachment until the ingredients are thoroughly combined. Add the wine, mixing until incorporated. Cover and refrigerate for at least 8 hours, and up to 24 hours.

**4.** Dissolve the Bactoferm in the distilled water. Mix the meat again, using the paddle attachment, adding the culture as you do so and mixing until the starter culture is well distributed and the meat is tacky.

**5.** Tie one end of each casing using a bubble knot (see page 122). Stuff the sausage into the casings, and tie each one off using another bubble knot. Using a clean needle, sausage pricker, or knife tip, poke holes all over the sausage, especially where there may be air pockets. If you're using weight to determine doneness, weigh the sausages and record the results.

**6.** Put the sausages on a baking sheet lined with parchment paper. Cover with another sheet of parchment, put another baking sheet on top, and weight it down with 8 to 10 pounds (4000 to 5000 grams) of weights. Refrigerate for 3 days.

**7.** Hang the sausages to incubate for about 24 hours in a warm place (ideally 80 degrees F./29 degrees C. and 80% humidity).

**8.** Hang the sausages in the drying chamber. If using the mold culture, mist them according to the package instructions. The salami are ready when they have lost about 30 percent of their raw weight, about 3 weeks.

**Yield: Two 1.75-pound/800-gram salami**

# Salami Felino

Felino is a town in the province of Parma in the heart of Emilia-Romagna. Hogs are said to have populated this area since the Bronze Age, and as the prosciutto of Parma became famous, so too did its salami. It's a simple sausage, dependent on the quality of the pork, seasoned with just salt, pepper, garlic, red wine, and a small amount of sugar for sweetness.

**2 pounds/900 grams pork shoulder butt, cut into large dice**
**2 pounds/900 grams boneless lean beef (chuck or round), cut into large dice**
**1 pound/450 grams pork back fat, cut into large dice**
**2 ounces/56 grams sea salt**
**1 teaspoon/7 grams DQ Curing Salt #2**
**2 teaspoons/3 grams finely ground black pepper**
**1 tablespoon/11 grams dextrose**
**3 garlic cloves, minced**
**½ cup/125 milliliters chilled dry red wine**
**1 tablespoon/10 grams Bactoferm (live starter culture; see Sources, page 267)**
**2 tablespoons/30 milliliters distilled water**
**Mold 600 (see page 70 for more information and page 267 for sources) (optional)**

**Two 18-inch/45-centimeter lengths beef middle, soaked in tepid water for at least 20 minutes and rinsed**

1. Partially freeze the meat and fat.

2. Grind the pork through a ¼-inch/6-millimeter (medium) die into the bowl of a stand mixer. Grind the beef into the bowl, then grind in the fat.

3. Add the salt, curing salt, pepper, dextrose, and garlic and mix on medium speed with the paddle attachment until the ingredients are well distributed. Add the wine and mix until incorporated.

*continued on next page*

**4.** Dissolve the Bactoferm culture in the distilled water, add it to the meat, and mix until it's well distributed and the meat is tacky, another minute or so.

**5.** Tie one end of each casing using a bubble knot (see page 122). Stuff the sausage into the casings, and tie each one off using another bubble knot. Using a clean needle, sausage pricker, or knife tip, poke holes all over the sausage, especially where there may be air pockets. If you're using weight to determine doneness, weigh the sausages and record the result.

**6.** Hang the sausage to incubate for 12 hours in a warm place (ideally 80 degrees F./ 27 degrees C. and 80% humidity).

**7.** Hang the sausages in the drying chamber. If using the mold culture, mist them according to the package instructions. The salami are ready when they have lost about 30 percent of their raw weight, about 3 weeks.

**Yield: Two 1.75-pound/800-gram salami**

# Salami Calabrese

These are spicy sausages in the style of Calabria, in southern Italy, with chile peppers and aromatic spices. They're shorter links, pressed so that they're oval, and have a shorter cure time. Our colleague Emilia Juocys, who helped test many of these recipes, took this on and worked really hard to get it right. It never seemed distinctive enough to her until she added the cinnamon. The sweet spice made all the other flavors click into place.

For even more flavor, use pancetta fat in place of some of the back fat.

**4.5 pounds/2040 grams pork shoulder butt, cut into large dice**
**8 ounces/225 grams pork back fat, cut into large dice**
**2 ounces/56 grams sea salt**
**1 teaspoon/7 grams DQ Curing Salt #2**
**2 teaspoons/4 grams cracked black pepper**
**1 teaspoon/1 gram red pepper flakes**
**½ teaspoon/1 gram ground cinnamon**
**¼ cup/65 milliliters chilled dry red wine**
**1 tablespoon/10 grams Bactoferm (live starter culture; see Sources, page 267)**
**2 tablespoons/30 milliliters distilled water**
**Mold 600 (see page 70 for more information and page 267 for sources) (optional)**

**10 feet/3 meters hog casings, soaked for at least 20 minutes in tepid water and rinsed**

**1.** Partially freeze the meat and fat.

**2.** Combine the meat, fat, salt, curing salt, pepper, red pepper flakes, and cinnamon and grind through a ⅜-inch/9-millimeter (large) die into the bowl of a stand mixer. Using the paddle attachment, mix on medium speed until all the ingredients are evenly distributed, about 30 seconds or so. Add the wine and mix until incorporated.

*continued on next page*

**3.** Dissolve the Bactoferm in the distilled water, add it to the meat, and mix until well distributed and the meat is tacky, another minute or so.

**4.** Stuff the sausage into the hog casings and tie off into 5-inch/12-centimeter links. If you're using weight to determine doneness, weigh the sausages and record the results.

**5.** Put the sausages on a baking sheet lined with parchment paper. Cover with another sheet of paper, put another baking sheet on top, and weight it down with 10 pounds/4000 grams of weights. Refrigerate for 3 days to give the sausages an oval shape.

**6.** Hang the sausages to incubate for about 24 hours in a warm place (ideally 80 degrees F./27 degrees C. and 80% humidity).

**7.** Hang the sausages in the drying chamber. If using the mold culture, mist them according to the package instructions. The salami are ready when they have lost about 30 percent of their raw weight, about 3 weeks.

**Yield: 3.5 pounds/1500 grams salami; about 20 links**

## Crespone

Crespone are salami made of meat stuffed into irregular casings from the end of the animal's digestive tract. If you want to make this in the traditional fashion, we'd suggest using a hog bung, but it will cure perfectly in a beef middle as well. This traditional salami is associated with Agrate Brianza, a small town just outside Milan in the Lombardy region. Pork shoulder and belly are mixed with white wine, garlic, and sweet spices (ginger, nutmeg, cloves, coriander, mace, cinnamon) that give the sausage a rich red color and a delicately sweet flavor. This is a wonderfully aromatic sausage, and it is soft, almost spreadable, due to the high fat content. Crespone is sometimes smoked before being air-dried, so if you'd like to cold-smoke it first, don't hesitate to do so (see page 75).

3 pounds/1360 grams pork shoulder butt, cut into large dice
2 pounds/900 grams pork belly, cut into large dice
2 ounces/56 grams sea salt
1 teaspoon/7 grams DQ Curing Salt #2
1 tablespoon/6 grams finely ground black pepper
1 ½ teaspoons/2 grams coriander seeds, toasted and ground
1 teaspoons/2 grams ground ginger
1 teaspoon/2 grams freshly grated nutmeg
½ teaspoon/1 gram ground cloves
½ teaspoon/1 gram ground mace
½ teaspoon/1 gram ground cinnamon
3 garlic cloves, minced
¼ cup/65 milliliters dry white wine
1 tablespoon/10 grams Bactoferm (live starter culture; see Sources, page 267)
2 tablespoons/30 milliliters distilled water
Mold 600 (see page 70 for more information and page 267 for sources) (optional)

1 hog bung (see Sources, page 267), rinsed, soaked for several days in several changes of water, and rinsed again

*continued on next page*

**1.** Partially freeze the meat and fat.

**2.** Combine the pork shoulder and belly and grind through a ¼-inch/6-millimeter (medium) die into the bowl of a stand mixer. Using the paddle attachment, mix on medium speed for 30 seconds or so. Add the salt, curing salt, pepper, coriander, ginger, nutmeg, cloves, mace, cinnamon, and garlic and mix until all the ingredients are well distributed, another minute or so. Add the wine and mix until incorporated.

**3.** Dissolve the Bactoferm in the distilled water, add to the meat, and mix until well distributed and the meat is tacky, another minute or so.

**4.** Stuff the sausage into the casings (hog bungs have only one opening) and tie off with a bubble knot (see page 122). Using a clean needle, sausage pricker, or knife tip poke holes all over, especially where there may be air pockets. If you're using weight to determine doneness, weigh it and record the results.

**5.** Allow the sausages to incubate for 12 hours in a warm place (ideally 80 degrees F./ 27 degrees C. and 80% humidity).

**6.** Hang the sausages in the drying chamber. If using the mold culture, mist the sausages according to the package instructions. The salami are ready when they have lost about 30 percent of their raw weight, about 3 weeks.

**Yield: Two 1.75-pound/800-gram or one 3.5-pound/1600-gram salami**

# Salami Diablo

The inspiration for this sausage came from our friend Marc Buzzio of Salumeria Biellese in New York City, one of the best salumi makers in the country. As the name indicates, it is a spicy sausage—use it when you want an aggressively spiced salami with some heat.

**3.5 pounds/1590 grams pork shoulder butt, cut into large dice**
**1.5 pounds/680 grams pork back fat, cut into large dice**
**2 ounces/56 grams sea salt**
**1 teaspoon/7 grams DQ Curing Salt #2**
**2 tablespoons/15 grams red pepper flakes, finely ground, plus 1 tablespoon/7 grams red pepper flakes**
**2 teaspoons/6 grams finely ground black pepper**
**1 tablespoon/8 grams fennel seeds**
**3 garlic cloves, minced**
**½ cup/125 milliliters chilled dry red wine**
**1 tablespoon/10 grams Bactoferm (live starter culture; see Sources, page 267)**
**2 tablespoons/30 milliliters distilled water**

**Two 18-inch/45-centimeter lengths beef middle, soaked in tepid water for at least 20 minutes and rinsed**

**1.** Partially freeze the meat and fat.

**2.** Grind the pork through a ¼-inch/6-millimeter (medium) die into the bowl of a stand mixer. Grind the fat into the bowl. Add the salt, curing salt, red pepper flakes, black pepper, fennel seeds, and garlic and, using the paddle attachment, mix on medium speed until well distributed, about 30 seconds. Add the wine and paddle until incorporated.

*continued on next page*

**3.** Dissolve the Bactoferm in the distilled water, add to the meat, and mix until well distributed and the meat is tacky, another minute or so.

**4.** Tie one end of each casing using a bubble knot (see page 122). Stuff the sausage into the casings, and tie each one off using another bubble knot. Using a clean needle, sausage pricker, or knife tip, poke holes all over the sausage, especially where there may be air pockets. If you're using weight to determine doneness, weigh the sausages and record the result.

**5.** Allow the sausages to incubate for 12 hours in a warm place (ideally 80 degrees F./ 27 degrees C. and 80% humidity).

**6.** Hang the sausages in the drying chamber. The salami are ready when they have lost about 30 percent of their raw weight, about 3 weeks.

**Yield: Two 1.75-pound/800-gram salami**

# Salami Nocciola

As the name indicates, this sausage features hazelnuts (*nocciola*), which add texture and crunch. Toasted pine nuts would also work well in place of the hazelnuts. Tarragon gives the sausage its aromatic depth.

**4 pounds/1815 grams pork shoulder butt, cut into large dice**
**1 pound/450 grams pork back fat, cut into large dice**
**4 garlic cloves, minced**
**2 ounces/56 grams sea salt**
**1 teaspoon/7 grams DQ Curing Salt #2**
**1 tablespoon/6 grams finely ground white pepper**
**1 tablespoon/12 grams dextrose**
**1 tablespoon/4 grams chopped fresh tarragon or 1 teaspoon/1 gram dried tarragon**
**¼ cup/65 milliliters chilled dry red wine**
**1 tablespoon/10 grams Bactoferm (live starter culture; see Sources, page 267)**
**2 tablespoons/30 milliliters distilled water**
**1 cup/120 grams hazelnuts, toasted, skinned, and coarsely chopped**
**Mold 600 (see page 70 for more information and page 267 for sources) (optional)**

**Two 18-inch/45-centimeter lengths beef middle, soaked in tepid water for at least 20 minutes and rinsed**

**1.** Partially freeze the meat and fat.

**2.** Combine the pork, fat, and garlic and grind through a ¼-inch/6-millimeter (medium) die into the bowl of a stand mixer.

**3.** Add the salt, curing salt, white pepper, dextrose, and tarragon and, using the paddle attachment, mix on medium speed until the ingredients are well distributed, 30 seconds or so. Add the wine and mix until the ingredients are distributed.

*continued on next page*

**4.** Dissolve the Bactoferm in the distilled water, add it to the meat, and mix until thoroughly incorporated. Add the hazelnuts and mix until combined and the meat is tacky.

**5.** Tie one end of each casing using a bubble knot (see page 122). Stuff the sausage into the casings, and tie each one off using another bubble knot. Using a clean needle, sausage pricker, or knife tip, poke holes all over the sausage, especially where there may be air pockets. If you're using weight to determine doneness, weigh the sausages and record the results.

**6.** Allow the sausage to incubate for 12 hours in a warm place (ideally 80 degrees F./ 27 degrees C. and 80% humidity).

**7.** Hang the sausages in the drying chamber. If using the mold culture, mist them according to the package instructions. The salami are ready when they have lost about 30 percent of their raw weight, about 3 weeks.

**Yield: Two 1.75-pound/800-gram salami**

## Salsicca di Cinghiale Crudo

*Cinghiale* is wild boar in Italian, and boar makes fabulous sausage. These sausages are popular wherever wild boar are hunted. While we don't need to worry about trichinosis from domestic hogs these days, we do have to concern ourselves when working with animals that might pick up the parasites from other animals in the wild. We recommend that you freeze the meat for 3 weeks before making this sausage (if you have a very cold freezer, you can reduce this time; see Note below). But, there is nothing that compares to wild meat, especially wild boar. It has a rich flavor and deep color that comes from muscles that are well worked in the wild. If you don't hunt your own boar, free-range boar is available by mail-order (see Sources, page 267).

**3 pounds/1360 grams wild boar shoulder meat or trim, frozen (see Note below) and cut into large dice**
**1 pound/450 grams pork shoulder butt, cut into large dice**
**2 ounces/56 grams sea salt**
**1 teaspoon/7 grams DQ Curing Salt #2**
**1 tablespoon/6 grams coarsely ground black pepper**
**1 tablespoon/9 grams coriander seeds, toasted and ground**
**2 teaspoons/4 grams cayenne pepper**
**2 teaspoons/3 grams ground mace**
**1 tablespoon/10 grams dextrose**
**1 pound/450 grams pork back fat, cut into large dice and well chilled**
**½ cup/120 milliliters dry white wine**
**1 tablespoon/10 grams Bactoferm (live starter culture; see Sources, page 267)**
**2 tablespoons/30 milliliters distilled water**
**Mold 600 (see page 70 for more information and page 267 for sources) (optional)**

**Two 18-inch/45-centimeter lengths beef middle, soaked in tepid water for at least 20 minutes and rinsed**

*continued on next page*

**1.** Partially thaw boar and combine with the pork, salt, curing salt, pepper, coriander, cayenne, mace, and dextrose and grind through a ½-inch/12-millimeter (extra-large) die into a bowl. Cover and refrigerate for at least 8 hours, and up to 24 hours.

**2.** Add the fat to the ground meat and grind it through a ¼-inch/6-millimeter (medium) die into the bowl of a stand mixer. Using the paddle attachment, mix on medium speed, adding the wine as you do so, until the ingredients are well distributed, about 1 minute.

**3.** Dissolve the Bactoferm in the distilled water, add the meat, and mix until well incorporated and the meat is tacky, another minute or so.

**4.** Tie one end of each casing using a bubble knot (see page 122). Stuff the sausage into the casings, and tie each one off using another bubble knot. Using a clean needle, sausage pricker, or knife tip, poke holes all over the sausage, especially where there may be air pockets. If you're using weight to determine doneness, weigh the sausages and record the results.

**5.** Allow the sausages to incubate for 12 hours in a warm place (ideally 80 degrees F./ 27 degrees C. and 80% humidity).

**6.** Hang the sausages in the drying chamber. If using the mold culture, mist the sausages according to the package instructions. The salami are ready when they have lost about 30 percent of their raw weight, about 3 weeks.

**Yield: Two 1.75-pound/800-gram salami**

**Note**: Freeze the boar for 12 days at -10 degrees F./-23 degrees C. or for 6 days at -20 degrees F./-30 degrees C.

# Nduja di Calabria

This is among the most unusual preparations in the salumi repertoire—in effect, a spreadable salami. A specialty of southern Italy, it's dense with paprika and fiery hot. Often an Italian chile paste is used (Alfonso Esposito *concentrato di peperoncini rossi picanti* is one brand we've found), but here we use La Vera paprika because it's both widely available and excellent. This is not a recipe to skimp on with generic paprika. The sausage also relies on a high fat ratio for texture and flavor, so we use all pork belly. Its deep crimson color and thick spreadable texture have led some to call this "red Nutella."

The nduja is lightly smoked and hung to dry for 1 week. We think you'll have the best results using hog casing. To serve it, peel back the casing and spread it on freshly toasted crostini.

**5 pounds/2270 grams fresh pork belly, cut into large dice**
**2 ¼ cups/300 grams La Vera *picante* (hot) paprika**
**1 cup/140 grams La Vera *agridulce* (bittersweet) paprika**
**2 ounces/56 grams sea salt**
**1 tablespoon/8 grams dextrose**
**1 teaspoon/7 grams DQ Curing Salt #2**
**1 tablespoon/10 grams Bactoferm (live starter culture; see Sources, page 267)**
**2 tablespoons/30 milliliters distilled water**

**10 feet/3 meters hog casings, soaked in tepid water for at least 20 minutes and rinsed**

**1.** Partially freeze the pork belly.

**2.** Grind the pork belly through a ¼-inch/6-millimeter (medium) die, then grind it again through an ⅛-inch/3-millimeter (small) die into the bowl of a stand mixer. Add the paprika, salt, dextrose, and curing salt and mix on medium speed for about a minute.

*continued on next page*

3. Dissolve the starter culture in the distilled water, add it to the meat, and mix until well distributed.

4. If using beef middles, tie one end of each casing using a bubble knot (see page 122). Stuff the sausage into the casings, and tie each one off using another bubble knot. Using a clean needle, sausage pricker, or knife tip, poke holes all over the sausage, especially where there may be air pockets.

5. If using hog casings, tie off one end using a bubble knot (see page 122). Stuff the sausage into the casing and tie off into 5-inch/12-centimeter links.

6. Hang the sausages to incubate for 24 hours in a warm place (ideally 80 degrees F./ 29 degrees C. and 80% humidity).

7. Cold-smoke the sausages for 4 hours.

8. Hang the sausages in the drying chamber. If using beef middles, hang for 2 to 3 weeks; if using hog casings, hang for 1 week.

9. Refrigerate the sausages, wrapped in butcher's paper, for up to 1 month.

**Yield: Two 18-inch/45-centimeter or twenty 5-inch/12-centimeter nduji**

# Whole-Muscle Salumi

Curing whole muscles is perhaps the best way for a novice to enter the craft of dry curing. It's a much less complex system than salami and relies on the inherent flavor of the meat and fat. But because of this, it's even more important that you use pork that was well raised, meaning that it lived on a good and natural diet, was free to run around and root according to its nature, and was as humanely slaughtered as possible. This usually involves connecting with a local farmer, which is the best option. But if that isn't possible, good mail-order pork is available (see Sources, page 267).

In the same way that most of the recipes in the salami chapter are variations on our basic salame recipe, the recipes here are simply variations on the basic themes laid out in Part One. In addition to different seasoning combinations for those whole muscles, this chapter also includes non-pork preparations, dry cures for beef, lamb, and goose.

All the standard techniques and safety issues covered in Curing Whole Muscles, page 78, apply here. Please read that, as well as the rest of the About These Recipes section before pursuing the dry-cured whole muscle.

## Spicy Guanciale

If you want to give dry-cured hog jowl some heat, add chiles to both your cure and the dry rub. This, of course, will add heat not only to the jowl itself but to whatever you cook that jowl with. It works beautifully transformed into *spuma di gota* (see page 221).

*continued on next page*

THE CURE

6 ounces/170 grams sea salt (or 8.5% of the weight of the jowl)

1 tablespoon/9 grams black peppercorns, roughly cracked in a mortar with a pestle or beneath a sauté pan

2 teaspoons/2 grams red pepper flakes

3 garlic cloves, minced

One 4.5-pound/2000-gram hog jowl, skin on, all glands removed

Dry white wine for rinsing the meat (optional)

THE RUB

2 tablespoons/12 grams *picante* (hot) paprika, preferably La Vera

1 tablespoon/9 grams black peppercorns, roughly cracked in a mortar or beneath a sauté pan

1 teaspoon/1 gram red pepper flakes

**1.** Combine the cure ingredients. Slip the jowl into a 2.5-gallon/9-liter zip-top plastic bag and pour in the cure. Rub the jowl well with the cure and seal the bag.

**2.** Put the jowl on a baking sheet, put another pan on the jowl, and weight it down with about 8 pounds/3600 grams of weights. Refrigerate for 2 days.

**3.** Rub the salt and juices around the jowl to redistribute them, flip the jowl, and reweight it. Refrigerate for 2 more days.

**4.** Remove the jowl from the bag and rinse it under cold water, then pat dry. Rub with the wine if you wish.

**5.** Combine the ingredients for the rub and rub them evenly over the jowl.

**6.** Poke a hole through a corner of the jowl, run a piece of butcher's string through it, and knot it. Hang the jowl in the drying chamber for up to 3 to 5 weeks, until it has firmed up; it should be stiff but still have some give.

**Yield: One 3-pound/1300-gram jowl**

# Fennel-Cured Coppa

Wild fennel (*finocchio*) grows all over Tuscany. It's fabulously aromatic, with a heady, sweet fragrance and flavor, and it's among the most common seasonings used with coppa. If you can't find wild fennel, use the best-quality seeds you can get—it makes a difference (see Sources, page 268). We use about a tablespoon each of pepper and fennel for every 5 pounds/2270 grams or so in the cure and twice that for the aromatic coating—the coppa should be evenly dusted. And we use about 1 garlic clove per pound or 500 grams in the cure.

**THE CURE**

**Coarse sea salt or kosher salt**
**Black peppercorns, toasted and roughly cracked in a mortar with a pestle or beneath a sauté pan**
**Fennel seeds, toasted and roughly cracked in a mortar or beneath a sauté pan**
**Garlic cloves, sliced paper-thin**

**1 coppa**
**Dry white wine for rinsing the meat (optional)**

**THE AROMATICS**

**Black peppercorns, toasted and roughly cracked in a mortar or beneath a sauté pan**
**Fennel seeds, toasted and roughly cracked in a mortar or beneath a sauté pan**

**1.** Dredge the coppa in salt (see The Salt-Box Method, pages 79–80) and put it in a 2.5-gallon/9-liter zip-top plastic bag. Add the peppercorns, fennel seeds, and garlic to the bag, 1 or 2 tablespoons of each, or until it looks good. Mark the bag with the coppa's weight and the date. Squeeze as much air out of the bag as possible and seal the bag.

*continued on next page*

**2.** Put the coppa on a baking sheet. Put another pan on top of the coppa and weight it down with 8 pounds/3600 grams of weights. Refrigerate for 1 day per each 2 pounds/1000 grams. Flip the coppa midway through the curing, redistributing the salt and aromatics as you do so, and weight it again.

**3.** Remove the coppa from the bag and rinse it under cold water. Pat it dry with paper towels and rub it with the wine if you wish.

**4.** Combine the peppercorns and fennel seeds and dust the coppa with them, evenly coating all surfaces.

**5.** If you're using weight to determine doneness, weigh the meat and record the result. Tie the coppa as you would a roast (see the illustrations on pages 122–25) and hang in the drying chamber for 4 to 6 weeks, or until it has lost 30 percent of its raw weight.

**Yield: 1 coppa**

# Coppa Cured with Bay Leaf and Juniper

The bay and juniper add sweet-savory complexity to this coppa. Juniper berries, with their wonderful gin smell, are great aromatics for pancetta and guanciale as well (find them at spice markets or online; see Sources, page 268). We like the intensity of dried herbs (provided they're fresh; see page 83 for more on using dried herbs). We use a couple tablespoons each of black pepper, juniper berries, thyme, and bay leaves, judging by eye.

**THE CURE**

**Coarse sea salt or kosher salt**
**Black peppercorns, toasted and roughly cracked in a mortar with a pestle or beneath a sauté pan**
**Juniper berries, coarsely ground or chopped**
**Dried thyme**
**Ground dried bay leaves**

**1 coppa**
**Dry white wine for rinsing the meat (optional)**

**THE AROMATICS**

**Black peppercorns, toasted and finely ground**
**Juniper berries, coarsely ground**
**Dried thyme**
**Ground dried bay leaves**

**1.** Dredge the coppa in salt (see The Salt-Box Method, pages 79–80) and put it in a 2.5-gallon/9-liter zip-top plastic bag. Add the peppercorns, juniper berries, thyme, and bay leaves to the bag, 1 or 2 tablespoons of each, or until it looks good. Mark the bag with the date. Squeeze as much air out of the bag as possible and seal the bag.

*continued on next page*

2. Put the coppa on a baking sheet. Put another pan on top of the coppa and weight it down with 8 pounds/3600 grams of weights. Refrigerate for 1 day per each 2 pounds/1000 grams. Flip the coppa midway through the curing, redistributing the salt and aromatics as you do so, and weight it again.

3. Remove the coppa from the bag and rinse it under cold water. Pat it dry with paper towels, and rub it with the wine if you wish.

4. Combine the aromatics and dust the coppa with them, evenly coating all surfaces.

5. If you're using weight to determine doneness, weigh the meat and record the result. Tie the coppa as you would a roast (see the illustrations on pages 122–25) and hang in the drying chamber for 4 to 6 weeks, or until it has lost 30 percent of its raw weight.

**Yield: 1 coppa**

# Spicy Coppa

Mild, sweet, porky coppa is delicious when you give it some heat. You could simply use cayenne, but we like to use peppercorns and paprika to give some fruitiness to the heat. We use about 3 tablespoons of black pepper for every 5 pounds/2270 grams of meat in the cure, and for the aromatics, we use about 2 tablespoons cracked black pepper, 2 tablespoons La Vera *agridulce* (bittersweet) paprika, 2 tablespoons La Vera *picante* paprika, and 2 teaspoons cayenne pepper.

**THE CURE**

**Coarse sea salt or kosher salt**

**Black peppercorns, toasted and roughly cracked in a mortar with a pestle or beneath a sauté pan**

**One 5-pound/2270-gram coppa**

**Dry white wine for rinsing the meat (optional)**

**THE AROMATICS**

**Black peppercorns, toasted and roughly cracked in a mortar or beneath a sauté pan**

***Agridulce* (bittersweet) paprika, preferably La Vera**

***Picante* (hot) paprika, preferably La Vera**

**Cayenne pepper**

**1.** Dredge the coppa in salt (see The Salt-Box Method, pages 79–80) and put it in a 2.5-gallon/9-liter zip-top plastic bag. Add the peppercorns to the bag, 1 or 2 tablespoons, or until it looks good. Mark the bag with the coppa's weight and the date. Squeeze as much air out of the bag as possible and seal the bag.

**2.** Put the coppa on a baking sheet. Put another pan on top of the coppa and weight it down with 8 pounds/3600 grams of weights. Refrigerate for 1 day per each 2 pounds/1000 grams. Flip the coppa midway through the curing, redistributing the salt and pepper as you do so, and weight it again.

*continued on next page*

**3.** Remove the coppa from the bag, and rinse it under cold water. Pat it dry with paper towels and rub it with the wine if you wish.

**4.** Combine the aromatics and dust the coppa with them, evenly coating all surfaces.

**5.** If you're using weight to determine doneness, weigh the meat and record the result. Tie the coppa as you would a roast (see the illustrations on pages 122–25) and hang in the drying chamber for 4 to 6 weeks, or until it has lost 30 percent of its raw weight.

**Yield: 1 coppa**

# Lonza Cured with Orange and Fennel

Orange and fennel are one of the most perfect pairings, and they work beautifully to flavor dry-cured meats and salami. The felicitous nature of the pairing is especially evident with the lean loin, which makes a fine platform for these flavors.

**Boneless pork loin, heavy sinew removed, with some of the back fat left on if you wish**

**THE CURE**

**Coarse sea salt or kosher salt**

**Fennel seeds, toasted and roughly cracked in a mortar with a pestle or beneath a sauté pan**

**2 oranges, sliced as thin as possible (including peel and seeds)**

**Juice of 1 orange**

**8 garlic cloves, sliced paper-thin**

**Dry white wine for rinsing the meat (optional)**

**Fennel seeds, toasted and finely ground**

**1.** Weigh the pork loin and record the result. Dredge it in salt (see The Salt-Box Method, pages 79–80) and put it in a 2.5-gallon/9-liter zip-top plastic bag. Add the fennel seeds, oranges, orange juice, and garlic to the bag. Squeeze as much air from the bag as possible and seal the bag.

**2.** Put the loin on a baking sheet. Put another pan on top and weight it down with 8 pounds/3600 grams of weights. Refrigerate for 1 day per every 2 pounds/1000 grams. Flip the coppa midway through the curing, redistributing the salt and aromatics as you do so, and weight it again.

**3.** Remove the loin from the bag, rinse under cold water, and pat dry. Rub with the wine if you wish.

*continued on next page*

**4.** Dust the loin with the ground fennel, lightly coating all surfaces.

**5.** Tie the loin as you would a roast (see the illustrations on pages 122–25). If you're using weight to determine doneness, weigh the meat and record the result. Hang in the drying chamber for 3 to 4 weeks, or until it has lost 30 percent of its raw weight.

**Yield: 1 lonza**

## Filetto

*Filetto* is Italian for tenderloin, and it's very easy to cure because it's small and lean, shortening the curing time and lessening the danger of rancidity. Here we flavor it with pepper and fennel. This recipes cures two tenderloins; halve the cure and aromatic ingredients if you want to cure only one.

**THE CURE**

**1.25 ounces/34 grams sea salt (salt-box method, see pages 79–80, or 3% of the weight of the meat)**
**1 tablespoon/6 grams black peppercorns, toasted and roughly cracked in a mortar with a pestle or beneath a sauté pan**
**1 tablespoon/6 grams fennel seeds, toasted and roughly cracked in a mortar or beneath a sauté pan**
**5 garlic cloves, sliced paper-thin**

**Two 3.5-pound/1500-gram pork tenderloins**
**Dry white wine for rinsing the meat (optional)**

**THE AROMATICS**

**2 tablespoons/12 grams black peppercorns, toasted and finely ground**
**2 tablespoons/12 grams fennel seeds, toasted and finely ground**

1. Combine the cure ingredients in a 2.5-gallon/9-liter zip-top plastic bag. Put the tenderloins in the bag and rub the cure all over them. Squeeze as much air out of the bag as possible and seal the bag. Refrigerate the tenderloins for 2 days, turning the bag halfway through to redistribute the cure.

2. Remove the tenderloins from the bag, rinse under cold water, and pat dry. Rub with the wine if you wish.

3. Combine the ground peppercorns and fennel and dust the tenderloins with them, coating all surfaces.

4. Tie the loins as you would a roast (see the illustrations on pages 122–25). If you're using weight to determine doneness, weigh the meat and record the result. Hang in the drying chamber for 3 to 4 weeks, or until the tenderloins have lost 30 percent of their raw weight.

**Yield: Two 2.5-pound/1135-gram filetti**

# Paletta Cruda

In the small town of Coggiola in Piedmont, east of Biella, we came across this interesting regional ham. It's a boned whole shoulder, cured, stuffed into a hog's bladder with local aromatics, and then either completely dried or partially dried and cooked. Dried, it is sliced and served like prosciutto; cooked, it is served sliced, like cooked ham. We give both methods here. We use pink salt, sodium nitrite, in the cure to prevent the growth of botulism bacteria. The salt also keeps the cooked paletta pink and gives it a hammy flavor.

Hog's bladder, sadly, is not available in the United States, but there are acceptable alternatives (see page 84 for a description, and see Sources, page 267). The bladder itself isn't critical to the preparation; it simply acts as a protective barrier, a hanging vessel that would otherwise go unused.

1 pork shoulder, skin removed, boned, and trimmed
Dry white wine for rinsing the meat (optional)
1 laminated hog's bladder, soaked overnight in cold water until pliable (see Sources, page 267)

THE CURE

7 ounces/200 grams sea salt (or about 8.5% of the weight of the meat)
1 teaspoon/7 grams pink salt

THE AROMATICS

1 tablespoon/8 grams black peppercorns toasted, roughly cracked in a mortar or beneath a sauté pan
1 teaspoon/3 grams coriander, toasted, roughly cracked in a mortar or beneath a sauté pan
1 teaspoon/1.5 grams finely ground dried thyme
1 teaspoon/1.5 grams finely ground dried rosemary

**1.** Weigh the meat and record the result. Combine the salt and pink salt in a 2.5-gallon/9-liter zip-top plastic bag. Give it a shake to distribute the pink salt. Add the meat and rub the cure all over it. Squeeze as much air as possible from the bag and seal the bag.

**2.** Refrigerate the meat for 1 day per each 2 pounds/1000 grams. Flip the meat midway through the curing, redistributing the salt as you do so.

**3.** Remove the meat from the bag, rinse under cold running water, and pat dry. Rub with the wine if you wish.

**4.** Mix the aromatics together. Dust the cured shoulder evenly with them, coating all surfaces. Stuff the shoulder into the bladder and sew it up in the same fashion as for the culatello (see illustrations 6–8 on page 177). Then tie it in the same manner as the culatello (see illustrations 9 and 10 on page 177), weighing it if using weight to determine doneness.

**5.** Hang the paletta in the drying chamber for 3 to 6 months, until it's lost 30 percent of its raw weight.

**Yield: One 6½-pound/3000-gram paletta**

## Paletta Cotto

This is dried for only 1 month, then cooked. Serve it at room temperature, thinly sliced.

Follow steps 1 through 4, and hang the paletta to dry for 1 month. Then simmer it in water to cover for 2 hours. Drain, return it to the pot, cover with fresh water, and simmer again for 2 hours. Allow to cool completely. Remove the bladder before serving.

# Speck

Smoked salumi is common in northeastern Italy, which retains many of the culinary customs that were the result of the Austrian occupation. The country bacon of the Alto-Adige region, for instance, is speck, smoked dry-cured ham. Ours is heavily seasoned with pepper, bay leaf, juniper, nutmeg, and cinnamon. Traditionally speck hams were moved to hang in the chimneys of people's houses in the spring, when fires were no longer in use, from which they picked up the desired smoke flavor. Now they're cold-smoked, as this one is (see page 75). Speck is usually the whole back ham of the hog, but here we use the shoulder, which is a smaller and more manageable cut. We love the aromatic impact of the juniper here. If you have access to juniper bushes, the branches can be used for smoking it.

**One 10-pound/4500-gram boneless pork shoulder (see pages 40–41 for how to bone this cut)**

**THE CURE**

**8 ounces/225 grams coarse sea salt or kosher salt (8.5% of the weight of the meat)**
**2 teaspoons/14 grams DQ Curing Salt #1**
**1 teaspoon/3 grams Colman's dry mustard**
**1 teaspoon/3 grams freshly grated nutmeg**
**4 tablespoons/48 grams finely ground juniper berries**
**4 tablespoons/48 grams ground allspice**
**4 tablespoons/48 grams finely ground black pepper**
**½ cup/85 grams dark brown sugar**
**1 tablespoon/10 grams finely ground dried bay leaves**

**1.** Lay the meat out flat, skin side down, on the work surface. With a cast-iron pan or other heavy pan or a meat mallet, pound the meat to an even thickness of about 4 to 5 inches/10 to 12 centimeters.

**2.** Combine the cure ingredients and add to a 13-gallon/49-liter plastic bag. Put the meat in the bag and rub it thoroughly with the cure.

**3.** Put the shoulder on a baking sheet, top it with another pan, and put about 8 pounds/3600 grams of weights on top pan. Refrigerate the shoulder for 5 to 6 days. Flip the shoulder each day, with the cure, to redistribute it and again, reweight it.

**4.** Remove the ham shoulder from the bag, rinse it under cold water, and pat it dry with paper towels.

**5.** Cold-smoke the shoulder for 5 to 8 hours at 80 to 90 degrees F./25 to 30 degrees C.

**6.** Weigh the meat if using weight to determine doneness. Poke a hole through a corner of the shoulder, run a piece of butcher's string through it, and knot it. Hang the meat in the drying chamber for about 6 months, until it has lost about 30 percent of its raw weight.

**Yield: 1 smoked shoulder**

# Culatello

If you want to cure your own culatello, remember that, as with any other dried ham, it is the breed of hog; the handling of the hog before, during, and after slaughter; the type of salt; and the drying conditions that give each ham its own distinct flavor. You'll need a hog's bladder—or the laminated version available in America—as well as a butcher's needle and twine (see Sources, page 267) for closing the seam of the bladder. If you don't feel comfortable describing the cut to your butcher, show the butcher the illustrations on page 176.

**One 10-pound/4500-gram culatello cut and trimmed**
**Sea salt as needed**
**1 cup/250 milliliters dry white wine**
**1 laminated hog's bladder (see Sources, page 267), soaked overnight in water**

**1.** Tie the culatello as shown in illustration 5 on page 177. Salt the culatello and using the salt-box method (see pages 79–80) put it in a 2.5-gallon/9-liter zip-top plastic bag. Seal the bag squeezing out excess air as you do.

**2.** Refrigerate (ideally at 38 degrees F./3.3 degrees C.) for 1 day per each 2 pounds/1000 grams of culatello (6 to 7 days for a ham of this size), massaging it and rubbing it with the salt every other day. It can even go a day or two longer if you wish; given the low level of salt, you can't oversalt it.

**3.** Remove the meat from the bag, brush off as much salt as you can, and rinse it with the wine to wash off the remaining salt. Pat dry with paper towels.

**4.** Make a cut down one side of the hog's bladder almost to the bottom. Fold the bladder around the culatello and sew the seam shut, constantly keeping pressure on the bladder to remove any air pockets (see illustrations 6–8 on page 177). Tie the culatello with butcher's string (see illustrations 9 and 10 on page 177). This is tricky—take your time. The string is used to compact the ham as much as possible and to support it while it hangs. (Neatness doesn't affect flavor.) If you're using weight to determine doneness, weigh the culatello and record the result.

**5.** Hang the culatello in the drying chamber for at least 4 to 5 months, until the meat is dry yet still supple, or until it's lost 30 percent of its raw weight. (Under proper conditions, culatello can hang for 36 months or more.)

**Yield: 1 culatello**

## Fiocco

The fiocco is the smaller front half of the ham (see illustration 4 on page 176). Follow the instructions for the culatello, using salt in the amount of 8.5% of the weight of the ham and refrigerating it for 1 day for each 2 pounds/1000 grams of ham.

Often referred to as the king of prosciutto, culatello is the back half of the ham, cured in salt, massaged, sewn up in a hog's bladder, and dried for one to four years. It's a large pear-shaped collection of muscles that more or less forms the rump of the hog. The name derives from the word *culo*, Italian for ass (the thing you fall on, not the animal). The best known culatelli are made in Emilia-Romagna, near the town of Zibello on the Po River, by Massimo Spigaroli (see page 15) at his Antica Corte Pallavicini. The Spigarolis in effect wrested the culatello from death-by-commercialization-and-factory-production and put it back in the hands of the artisan producers—some of them, anyway. About 25,000 hams are cured in any given year, 5000 of which hang in Spigaroli drying chambers. Meat with the bone in tends to hang on to its moisture more resolutely than boneless meat. The ham removed from

*continued on page 178*

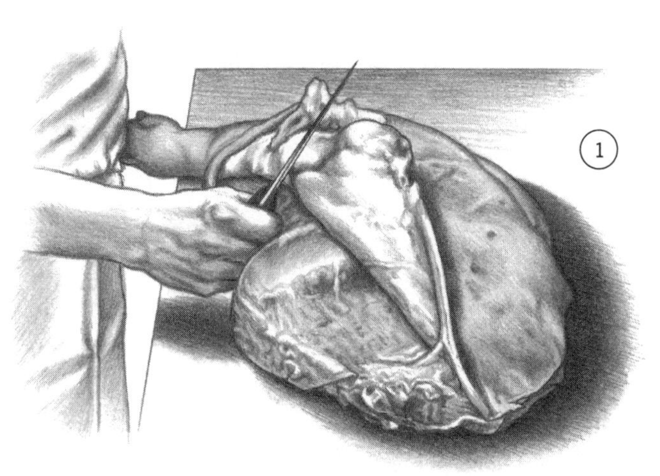


1. After scoring it, remove skin from the entire ham, being careful not to cut into the flesh. This takes some care and patience.

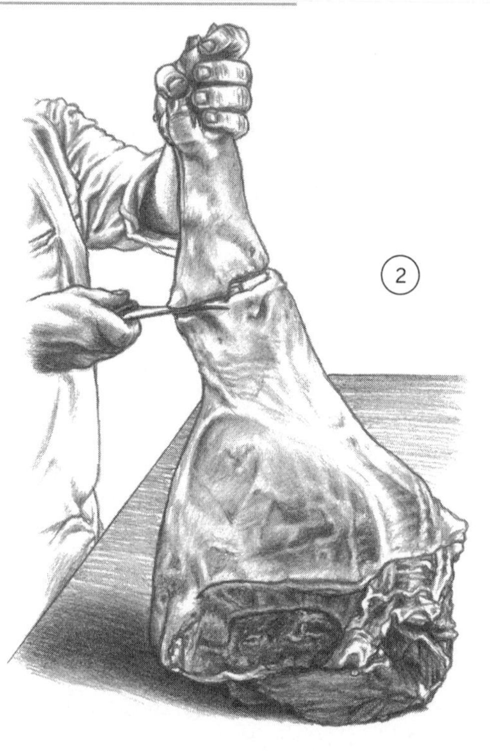


2. To begin the culatello cut, score the leg at the shank. This will give you your starting point in cutting the culatello and fiocco.

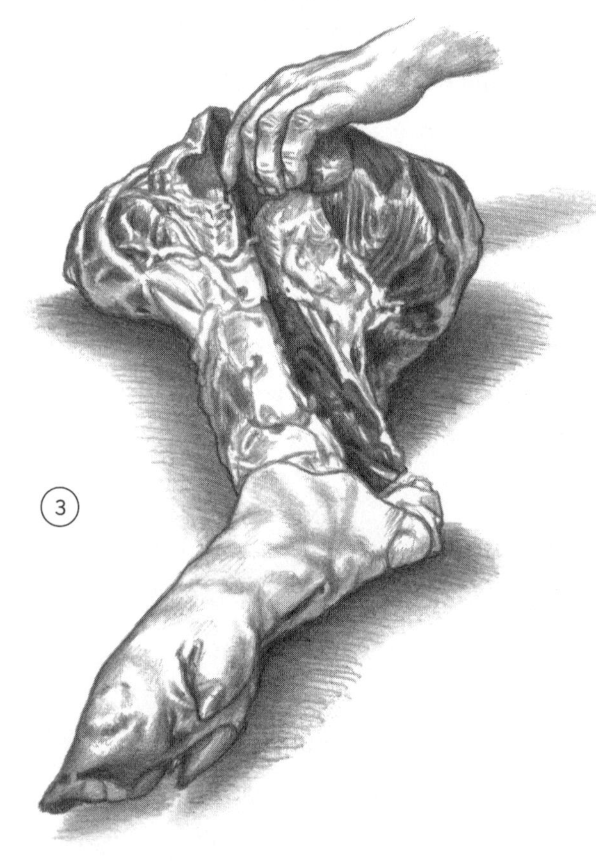


3. The beginning of the culatello cut. Starting at the scored joint, with your knife held perpendicular to the cutting board, slice the back part of the ham straight off the bone. This will be the culatello.

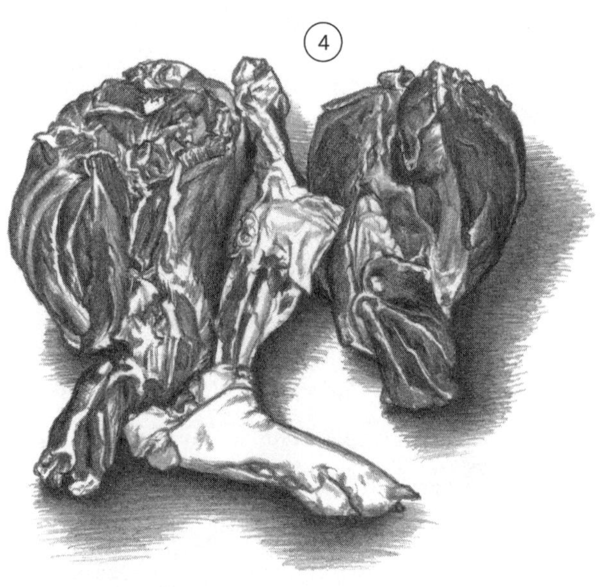


4. Here the leg has been flipped and the rest of the ham, the fiocco, has been removed. The culatello is on the left, the smaller fiocco on the right.

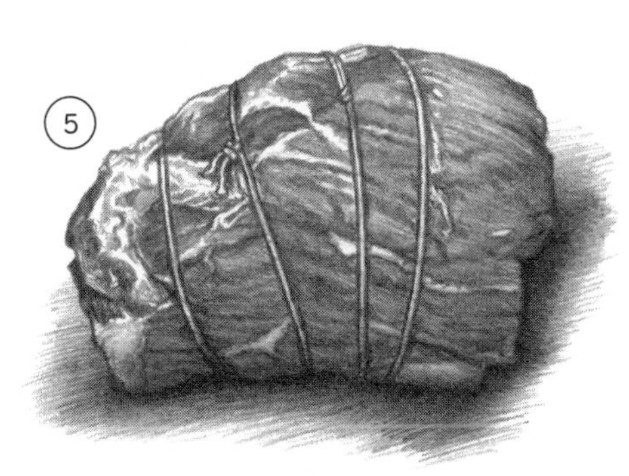


5. The culatello first needs to be shaped: tie it in 4 or 5 places. It's now ready to be salted.

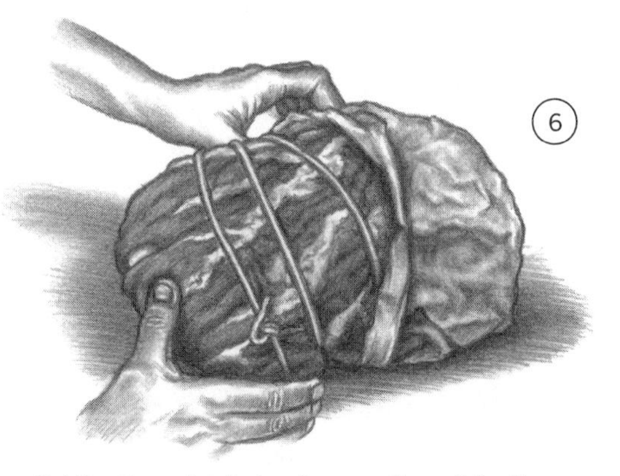


6. After the culatello has been on the salt for the requisite time, rinse and dry it and put it into the bladder (leaving the strings attached).

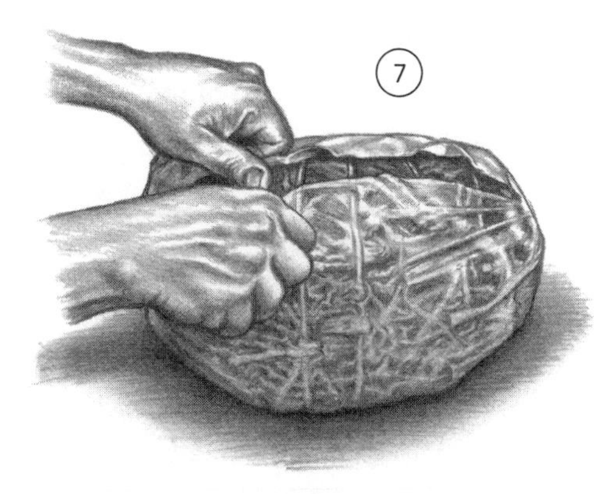


7. Pull the bladder tightly around the culatello to eliminate any air pockets.

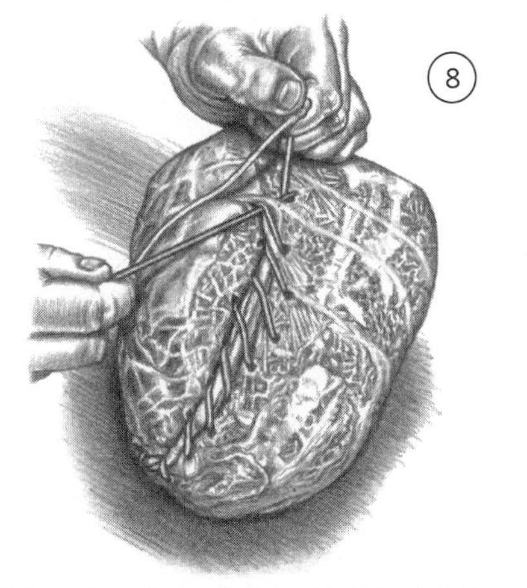


8. Using a trussing needle and butcher's twine, sew the bladder shut, starting at the small end and pressing out any air pockets in the process.

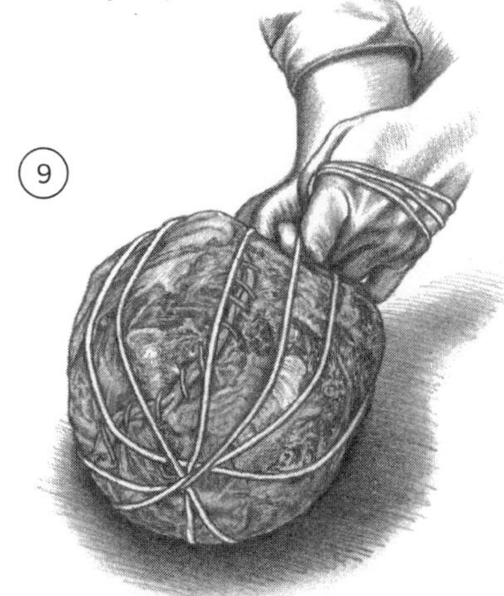


9. Here the culatello has been sewn up in the bladder. To make the support ties, loop the string around the culatello 5 times to create 10 vertical supports, crossing them at the bottom and top of the culatello, then tie off at the top.

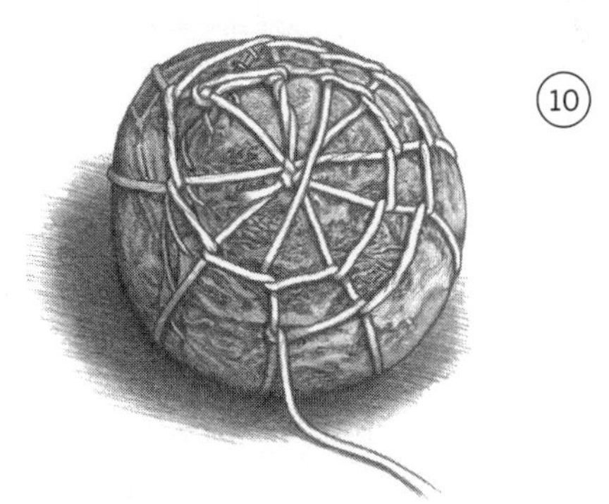


10. With another piece of string, secure the vertical supports at the bottom where they cross one another, and cut off 4 feet/1 meter of string. Loop the string under and around each vertical support, working your way around and up the culatello. Tie the string off at the top, leaving at least 18 inches/45 centimeters free to make a loop to hang it. The culatello will, from its own weight, form into a teardrop shape. (See page 16 for an illustration of finished culatellos.)

the bone, inhabitants of this part of Italy likely learned, cured beautifully here. We can attest to the truth of this.

The culatello in Zibello, both at Spigaroli's and at an inn not far from there, La Buca, was fabulous. (La Buca is a lovely little place, stereotypically wonderful: grandma, the proprietress, comes to your table, asks what you're in the mood for, and tells you what you're going to have. She cures her own salumi underneath the restaurant. She served us a big plate of circular tiles of paper-thin culatello, pale pink with beautiful marbling, and with it nothing more than bread and cold hard curls of butter.)

We had come to Italy in May, after the cold wet winter had passed, and toured the Spigaroli cellars in dry warmth. We took stone steps sided by stone walls down to the humid cellar, built in 1312. We ducked between thousands of moldy shrunken beauties, culatelli hanging like overripe tropical fruits from the wooden ceiling beams. The windows at the back were wide open to the river yards away. The people in charge of the cellar open and close the windows as is necessary to control moisture; the room usually has 85 percent humidity. The culatelli were shrinking out of their string ties, some dripping with dust-mite lace that fell from the hams like Spanish moss. Prominently displayed were culatelli hung for well-known personages, blackboards attached to them with names like Armani and Principe Carlos written on them. (Prince Charles sent ham from his own herd to the Spigarolis to cure for him.)

Later at the restaurant we ordered a tasting, and it was akin to a flight of wines, varying hams cured for varying times, all of them spectacularly tender and creamy, deeply flavorful, with the complex sweet porkiness of long-aged meat. Tasting the difference that aging makes enhances the sense of magic surrounding their creation. The final tasting, from a breed called Nera Parmigiana that had been cured for forty-five months—three months short of four years!—was so tender it almost liquefied in the mouth, and in addition to the usual aged flavors and aromas was the distinct rotty-moldy-fermented flavors of the finest blue cheese. But, really, there is nothing in our experience that is quite like it.

# Lamb Prosciutto with Garlic

Lamb prosciutto is a great variation on the classic dry-cured ham from both a flavor standpoint and the ease of the cure—the leg is boned, first of all, and it's smaller than a ham and so dries much more quickly. The simple cure of pepper, garlic, and salt results in a deeply lamby, garlicky finished meat. Serve this thinly sliced, as is, or in any way you would enjoy dried ham. (See the illustrations on pages 45–46 for carving prosciutto.)

**1 leg of lamb (about 8 pounds/3600 grams)**
**2 heads garlic, separated into cloves, peeled, and minced**
**½ cup/56 grams black peppercorns, toasted and coarsely ground**
**Coarse sea salt or kosher salt**
**Dry white wine for rinsing the leg (optional)**

**1.** Weigh the lamb and record the result. Rub the leg all over with the garlic, then coat it with the pepper.

**2.** Put the leg of lamb in a large plastic bag and add enough salt to cover it completely; it should be packed in salt. Seal the bag, squeezing out air as you do.

**3.** Put the lamb on a baking sheet, put another pan on top of it, and weight it with 8 pounds/4000 grams of weights. Refrigerate for 1 day for each 2 pounds/1000 grams it weighs. Inspect the lamb every day to make sure it's completely packed in salt.

**4.** Rinse the leg under cold water to remove all the salt. Rub it with the wine if you'd like.

**5.** Weigh the lamb if using weight to determine doneness. Tie the leg as you would a prosciutto and hang it in the drying chamber. The lamb is ready when it has lost about 30 percent of its raw weight, 3 to 4 months.

**Yield: 1 lamb prosciutto**

# Bresaola

Bresaola, dry-cured beef, is a specialty of the Lombardy region of Italy, originating in the Valtellina Valley. Many factors contribute to a great dried beef round: the quality of the beef, the cut of beef, the seasonings, and the drying process. But much in the same way that the microclimate around Zibello results in the excellence of the culatello dry-cured there, the alpine climate in which bresaola dries is what makes it unique.

You can dry cure any lean beef (beef fat is not great dried, so lean is important), but we prefer the eye of the round because of its uniform shape and size—it's perfect for dry curing.

While you can simply season it with pepper, we like to use an aromatic blend of spices. The salt is given as a percentage, but the spices are given as specific measurements based on about a 4-pound/1800-gram piece of meat. It can be lightly cold-smoked after curing if you wish for an additional level of complexity. The meat will be firm on the outside when it's done and silky-smooth inside when sliced. Serve it with slices of lemon, and give it a drizzle of good olive oil and a grinding of fresh black pepper.

**THE CURE**

**2 ounces/60 grams sea salt (salt-box method, see pages 79–80, or 3% of the weight of the meat)**
**2 teaspoons/7 grams freshly ground black pepper**
**2 teaspoons/6 grams fresh thyme leaves**
**10 juniper berries, crushed**
**1 teaspoon/1 gram finely ground dried bay leaves**
**1 teaspoon/1 gram ground cinnamon**
**1 teaspoon/1 gram ground cloves**
**½ cup/125 milliliters dry white wine**

**One 4-pound/2000-gram eye of the round roast**

**1.** Combine all the cure ingredients and add to a 2.5-gallon/9-liter zip-top plastic bag. Add the beef and rub it all over with the salt and seasonings. Seal the bag and refrigerate for 7 to 9 days, flipping it and rubbing it again every other day to make sure the cure is evenly distributed.

**2.** Remove the beef from the bag, rinse thoroughly under cold water to remove any remaining spices, and pat dry with paper towels. Set it on a rack on a baking sheet and let air-dry, uncovered, at room temperature for 2 to 3 hours. Weigh the beef and record the result.

**3.** With butcher's string, tie the beef to make a compact roast (see pages 122–25). Hang it in the drying chamber for about 3 weeks or until it's lost 30 percent of its weight.

**Yield: 2.5 pounds/1100 grams bresaola**

# Goose Prosciutto

By far one of the easiest cuts of meat to dry cure is a goose or duck breast. It's what we always recommend for first timers or anyone timid about dry curing. You don't even need a special dry-curing chamber, just a cool dry area away from the light. Any goose breast can be used for this recipe, from snow geese to Canadian geese. The breast of a duck raised in America for foie gras, called moulard, is almost as large as a goose breast and can be used here as well. Goose breast is little more full flavored than duck. The fat will be soft and creamy, and the meat will be a dark dense red with a distinctive cured flavor. To store it, wrap it in paper; it will keep for two weeks if properly dried.

**THE CURE**

**½ teaspoon/3 grams pink salt**

**½ teaspoon/1 gram finely ground dried bay leaves**

**½ teaspoon/1 gram finely ground dried thyme**

**4 juniper berries, crushed**

**5 black peppercorns, crushed**

**1 garlic clove, sliced paper-thin**

**Coarse sea salt or kosher salt**

**1 whole goose or duck breast, split, silverskin removed**

**1.** Combine the pink salt, bay leaves, thyme, juniper berries, and pepper in a small bowl and whisk together. Rub the garlic into the breasts and then dust with the seasoning mixture.

**2.** Put the breasts in a nonreactive pan just large enough to hold them, or in an appropriately sized plastic bag, and add enough salt so that they are completely encased in salt. Refrigerate for 24 hours.

**3.** Rinse the breasts under cold running water and pat dry. If you will be using weight to determine doneness, weigh the meat and record the result. Wrap the breasts in cheesecloth and tie in a continuous tie (see pages 122–25).

**4.** Hang the breasts in the drying area, for 1 to 3 weeks, or until they feel firm but are not hard. Refrigerate overnight before thinly slicing and serving.

**Yield: 2 goose or duck prosciutto halves**

# Cooked Salumi

Most salumi preparations don't involve heat, relying solely on salt and drying for their preservation, flavor, and texture. A few preparations do involve actual cooking though. The Soppressata di Tuscano, for instance, is a method created to make use of all the bits and parts of the hog that don't dry well: the head and ears, bones, trim. Cotechino (page 193) is a sausage that makes use of the skin, which is not typically dried on its own. The following recipes all involve cooking, and while some are not strictly considered salumi, they're all part of the salumi process.

## Soppressata di Tuscano/Coppa di Testa (Tuscan Headcheese)

In much of Italy, you'll see soppressata as a larger, dry-cured sausage, in the style of salami. But in Tuscany, ask for soppressata, and you'll be given a slice of mixed meats cooked like headcheese, often stuffed into an enormous casing. The preparation is borne of economy, a tasty way to make use of everything left over after all the other cuts are prepared: feet, tongue, head, and tail, cooked low and slow in a flavorful liquid that extracts enough gelatin to become sliceable when cooled, a binder for the assorted meats. It's exciting to see all different textures and shapes in the gel—a gorgeous mosaic of skin, ears, chunks of meat, and tongue (Brian likes to use two pork tongues). It's all deliciously flavored with wine and sweet spices and herbs that infuse the meat and stock. It can be sliced and served just like a pâté, with a salad, or as part of a salumi board.

You don't *have* to use the head; you could use shoulder instead. Just because this was created to make use of the head doesn't mean it won't work with any braising cut of the hog. Just be sure to use plenty of skin so you will have enough gelatin to set the stock into a sliceable form. Slow-cooking a variety of odd bits and setting them in flavorful gelled stock is a great all-purpose technique.

We chill this in a terrine mold, but you could also stuff it into a large natural or fabricated collagen casing, as they do in Tuscany. Another technique is to separate the solids, roll them into a cylinder in cheesecloth, put the cylinder into a terrine that holds it snugly, and pour the warm strained stock over it. The stock soaks into the cylinder and, as it sets, holds all the meat and tasty bits together, so that you can unroll it and slice perfect disks of headcheese for an elegant presentation.

1 hog's head (15 pounds/7720 grams or so)
2 hog's feet
2 smoked pork tongues or 1 smoked beef tongue (1.5 pounds/680 grams), peeled (optional)
8 thyme sprigs
8 parsley stems
4 bay leaves
1 teaspoon/3 grams black peppercorns, toasted and roughly cracked in a mortar with a pestle or beneath a sauté pan
1 cinnamon stick
3 whole cloves
20 allspice berries, toasted and cracked
½ teaspoon/1.5 grams coriander seeds, toasted and roughly cracked in a mortar with a pestle or beneath in a sauté pan
2 Spanish onions, cut into large dice
4 celery stalks, cut into large dice
4 carrots, cut into large dice
2 cups/470 milliliters dry Madeira
2 cups/470 milliliters dry white wine
Kosher salt

*continued on next page*

**1.** Rinse the head and feet under cold water. Put them in a large stockpot, cover them with water by 2 inches/5 centimeters, and bring to a boil. Drain and rinse well.

**2.** Return the head and feet to the clean pot, along with the tongue, if using. Tie up the thyme, parsley, bay leaves, peppercorns, cinnamon stick, cloves, allspice berries, and coriander seeds a cheesecloth bag, and add this *sachet d'épices* to the pot. Add the onions, celery, carrots, Madeira, and white wine, then add enough water to cover by 2 inches/5 centimeters. Bring to a simmer, skimming any foam, froth, or coagulated protein that rises to the surface, and cook until the meats are fork-tender. The tongue and hog's feet should be ready after 4 to 6 hours. Remove and let cool slightly. Pick the meat from the hog's feet and discard the skin, fat, and bones. Peel the tongue and cut into medium dice. Combine the meat in a bowl, pour a ladleful of stock over the meat to keep it moist, and cover with plastic wrap.

**3.** Remove the head when it is tender, about 6 hours. When it's cool enough to handle, pick off the meat and dice it. Remove the ears and slice into strips. Discard the skull, fat, and skin. Add the meat and ears to the bowl with the tongue and hog's feet, and add more hot liquid to cover.

**4.** Strain the stock through a colander, then strain again through a China cap/chinois lined with cheesecloth into a bowl or pot. Skim as much fat from the surface as possible. To check that you have enough gelatin, pour a small amount of stock onto a plate and chill in the refrigerator. When it's cold, the stock should be solid (usually there's plenty of gelatin from the hocks and skin to make a solid gel). If it's not, reduce the stock by about one-fifth and check again. When consistency is good evaluate for seasoning.

**5.** Line a 6-cup/1.5-liter terrine mold with plastic wrap, leaving an overhang on the two long sides. Arrange the meats loosely in the mold. Pour enough of the stock over to cover and fold the overhanging plastic over the top. Refrigerate for 8 to 12 hours, until set. The headcheese will keep in the refrigerator for up to a week.

**6.** To serve, remove headcheese from the mold and slice into ½-inch/1.25-centimeter slabs.

**Yield: Serves 10**

A good way to cool a Tuscan soppressata is to chill it in its poaching liquid or stock. The soppressata will soak up the gelatin-rich liquid, which will both flavor it and help it set into what is in effect a cylindrical headcheese.

# Mortadella

Mortadella, a specialty of Bologna, is what's called an emulsified sausage, in which the meat and fat are uniformly mixed. It should be as smooth and uniform as a slice of bologna. This version has an interior garnish of diced pork back fat for flavor, richness, and visual appeal.

8 ounces/225 grams boneless lean beef, cut into large dice
8 ounces/225 grams lean pork, cut into large dice
¼ cup/60 milliliters dry white wine
1 teaspoon/6 grams minced garlic
½ ounce/14 grams sea salt (about 1½ tablespoons)
12 ounces/340 grams pork back fat, cut into large dice
10 ounces/280 grams crushed ice
1½ teaspoons/4 grams finely ground black pepper
1 teaspoon/4 grams ground mace
1 teaspoon/2 grams coriander seeds, toasted and ground
½ teaspoon/1 gram ground dried bay leaves
½ teaspoon/1 gram freshly grated nutmeg
½ cup/120 grams diced (¼-inch/6-millimeter) pork back fat, blanched in boiling water for 1 minute, drained, and cooled

One 18-inch/45-centimeter length beef middle, soaked in tepid water for at least 20 minutes and rinsed

**1.** Combine the beef, pork, white wine, garlic, and salt and grind through an ⅛-inch/3-millimeter (small) die into a bowl set in ice; refrigerate until chilled, about an hour. Grind the large dice of fat through the same die into another bowl set in ice, and place in the freezer.

**2.** Place the ground meat in a food processor, along with the ice, pepper, mace, coriander, bay leaves, and nutmeg, and process until the mixture is smooth and has reached a temperature of 40 degrees F./4 degrees C., 4 to 5 minutes. Add the ground fat and process until the mixture reaches 45 degrees F./5 degrees C., 4 to 5 minutes.

**3.** Transfer the mixture to a bowl set in ice and fold in the blanched pork fat.

**4.** To check the seasoning, poach a spoonful of the mixture in simmering water for 3 to 4 minutes, until cooked through, then taste it and adjust the seasoning if you wish.

**5.** Tie one end of the casing using a bubble knot (see page 122). Stuff the sausage tightly into the casing and tie it off using another bubble knot.

**6.** Poach the sausage in 170-degree-F./76-degree-C. water until it reaches an internal temperature of 150 degrees F./65 degrees C, about 45 minutes to 1 hour. Transfer to an ice bath and chill thoroughly. Drain and refrigerate.

**Yield: One 3-pound/1500-gram mortadella**

# Zampone

This is one of the coolest hog dishes ever created. A hog shank, with the trotter and hoof left on, is boned out, keeping the skin intact; the leg is stuffed with a cotechino-like sausage filling and sewn up, slow-cooked, and then roasted slowly to a golden brown and served hoof and all. It's enormously festive and is commonly the centerpiece of a Christmas dinner. Legend has it that it was invented in Emilia-Romagna, in Modena, the balsamic vinegar capital, in 1511 for the troops of Pope Julius II, the warrior pope. The sausage is very rich, here with ears being included in the mix for their abundant gelatin and great texture (skin can be substituted if ears are hard to come by), and it pairs beautifully with lentils or another legume and a braised sturdy green such as kale or chard. Season with some great balsamic, in keeping with the story of its origins.

It takes some work and preparation, and you'll need a trussing needle and trussing thread, but for one of the greatest hog preparations ever, it's worth it.

**One 11-pound/5000-gram fresh front leg of pork, shoulder blade, trotter, and skin attached**
**2 fresh hog's ears**
**1 pound/450 grams well-marbled pork shoulder, cut into large dice**
**12 ounces/340 grams pork back fat**
**3 garlic cloves, minced**
**1 ounce/28 grams sea salt**
**½ teaspoon/3 grams pink salt (optional)**
**2 teaspoons/4 grams crushed black peppercorns**
**1 teaspoon/2 grams freshly grated nutmeg**
**1 teaspoon/2 grams ground cloves**
**½ teaspoon/1 gram finely ground dried bay leaves**
**½ teaspoon/1 gram cayenne pepper**

**GARNISHES**
**¼ cup/20 grams chopped fresh parsley**
**½ cup/60 grams pistachios, blanched briefly in boiling water and skinned**

**12 ounces/335 grams quartered mushrooms, sautéed in 2 tablespoons/30 grams butter, chilled**
**12 ounces/335 grams smoked ham, diced**

**1.** With the leg skin side down, make an incision in the inside of the leg at the top of the arm bone, starting where the blade bone connects to it (see the illustrations on page 41). Carefully remove the skin, without poking any holes in it and keeping it in one piece: cut through just the skin and carefully peel it back to the first joint of the trotter. Separate the arm bone from the trotters at the joint. You will have 2 pieces at this point, the skin with the trotter attached and the skinless bone-in spalla. Soak the skin and trotter in cool water for 2 hours, which will help soften the skin and make it more pliable.

**2.** Meanwhile, remove the blade bone, heavy sinew, and arm bone from the meat. Inspect it for glands, the brownish squishy nodules of what is clearly neither meat nor fat, cutting out and discarding any you find. (Save the shoulder blade and arm bone for stock.) Cut the meat into large dice and refrigerate.

**3.** Put the hog's ears in a medium saucepan, cover with 2 inches/5 centimeters of water, bring to a simmer, and simmer for 1 hour, or until tender. Drain and cool thoroughly under cold running water. Cut into ½-inch/1.25-centimeter pieces.

**4.** Combine the diced meat, ears, fat, garlic, salt, pink salt, if using, pepper, nutmeg, cloves, bay leaves, and cayenne and grind through a ¼-inch/6-millimeter (medium) die into the bowl of a stand mixer. Using the paddle attachment, mix for 2 minutes. Add the garnishes and mix for another minute. Cover with plastic wrap and refrigerate.

**5.** Lay the skin out on a work surface and, with a sharp knife, scrape the fat from the inside of the skin, trying to make it as evenly thin as possible, all the while being careful not to tear it. Trim the skin to make it a bit thinner if some areas are thick with fat or meat. With the X-Acto knife, box cutter, or razor blade, poke holes all over the skin (from the inside out). The holes will allow the skin to expand around the sausage filling as it expands with the heat and then to shrink as the filling cools and shrinks.

*continued on next page*

6. Trim the skin so you will be able to make a cylinder the same diameter as the trotter or slightly larger, getting wider toward where the blade used to be.

7. Shape and slap the sausage filling into a large meatball to ensure there are no air pockets, then form into a log the same length as the skin and place in the middle of the skin. Fold the skin over itself to enclose the sausage and sew it securely closed with trussing thread.

8. Put the sausage in a large pot, add stock or water to cover, and bring to a simmer. Poach at 170 degrees F./76 degrees C. for 2 hours, until the interior temperature reaches 150 degrees F./65 degrees C. Allow to cool in the poaching liquid, then refrigerate overnight.

9. Preheat the oven to 325 degrees F./162 degrees C.

10. Remove the zampone from the liquid, and pat dry. Cook for 45 minutes, or until heated through and golden brown. Slice to serve.

**Serves 8 to 10**

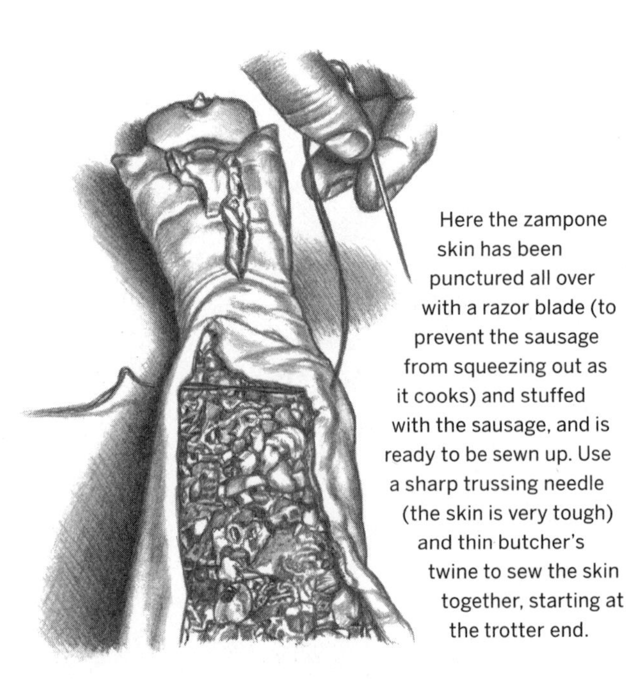

Here the zampone skin has been punctured all over with a razor blade (to prevent the sausage from squeezing out as it cooks) and stuffed with the sausage, and is ready to be sewn up. Use a sharp trussing needle (the skin is very tough) and thin butcher's twine to sew the skin together, starting at the trotter end.

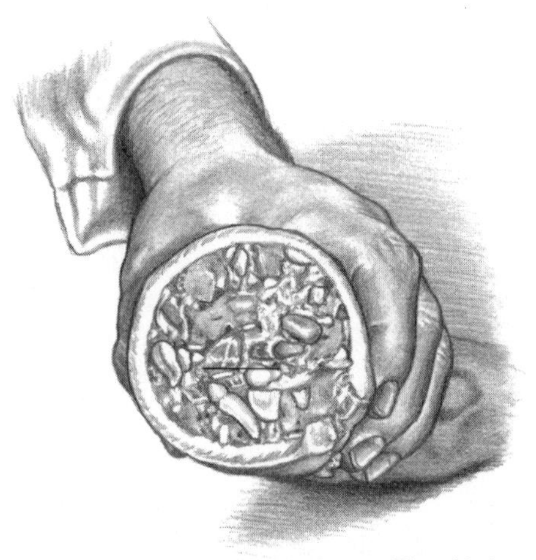

A cross section of the cooked zampone. The skin is soft and gelatinous, the filling is moist and studded with the garnish, ham, and pistachios.

Of all the big eight cuts the coppa is probably the easiest to cure successfully. It has excellent, intramuscular fat, is relatively narrow and so dries quickly, and accepts flavors well, whether aromatic fennel or spicy browned chiles.

Coppa

Soppressata (left) and Calabrese are two rustic salami distinctly different in flavor, but both are excellent examples of how to use the trim from whole-hog butchery.

Calabrese salami, coppa, lonza, and prosciutto (from top) with fresh white peaches, almonds, a plum, good bread, and dry red wine.

Speck, coppa, lonza, and Calabrese salami (from left) are served with fruit, crusty bread, and a Belgian endive marmalade (page 212).

A classic salumi plate (from left): Calabrese salami, lonza, speck, and prosciutto. Fresh white peaches and a plum are a sweet, juicy contrast to the salty salumi.

We sometimes serve fruity extra virgin olive oil, for the bread or to drizzle on the meat, here Calabrese salami, coppa, lonza, and prosciutto, along with Rainier and bing cherries.

Many different accompaniments go well with salumi. Here we have marinated artichokes, green olives, caper berries, roasted peppers, and fresh figs to complement Calabrese salami, prosciutto, lonza, speck, and coppa.

Prosciutto ready to be cut.

# Cotechino

The rind is sometimes removed before pancetta is rolled and hung to dry. What to do with this skin then? It's great to add to stocks and soups, because it releases copious gelatin that adds body. You can also poach it to tenderize it and then fry it up as cracklings, known in Italy as *cotiche*. Or you can feature this protein-rich goodness, either cured or fresh, in a sausage, cotechino. Sausage enriched with hog skin is one of the best you can make. The skin can be handled in differing ways: It can be added to the sausage mixture cooked or raw. It can be ground with the rest of the meat, or it can be sliced or chopped and added as an interior garnish.

The main thing to be aware of when working with skin is that it's naturally tough—it's connective tissue, protein. So you've got to tenderize it at some point. If it's added to the sausage mixture raw, the sausage must be cooked low and slow to tenderize the skin. If it's cooked before being added to the sausage, it doesn't need to be tenderized again. Because the tough skin is often added to sausage raw, cotechino is often braised with legumes (see Cotechino with Lentils on page 259), adding flavor and body to the stew as it cooks. Cotechino with lentils is often served on New Year's Day in Italy, not unlike our hoppin' John.

**1.5 pounds/680 grams pork skin**
**2.5 pounds/1135 grams pork shoulder butt, cut into large dice**
**1 pound/450 grams pork back fat, cut into large dice**
**2 ounces/56 grams sea salt**
**1 teaspoon/7 grams pink salt (optional)**
**2 teaspoons/6 grams coarsely ground black pepper**
**½ teaspoon/1 grams cayenne pepper**
**½ teaspoon/1 gram ground cinnamon**
**½ teaspoon/1 gram ground cloves**
**½ teaspoon/1 gram finely ground dried bay leaves**
**⅛ teaspoon/0.5 gram freshly grated nutmeg**
**⅛ teaspoon/0.5 gram ground mace**

*continued on next page*

3 garlic cloves, minced

½ cup/125 milliliters dry red wine

Two 18-inch/45-centimeter lengths beef middle, soaked in tepid water for at least 20 minutes and rinsed

About 3 quarts/3 liters stock or water

**1.** Cut the pork skin into 3-inch/7.5-centimeter squares. Put in a medium saucepan, cover with an inch/2.5 centimeters of water, and bring to a simmer, skimming off the foam that rises, until tender, about 1 hour. Drain and rinse the skin under cold water, then cover and refrigerate until thoroughly chilled.

**2.** Grind the pork skin through a ¼-inch/6-millimeter (medium) die into a bowl.

**3.** Add the pork, fat, salt, pink salt, pepper, cayenne, spices, and garlic and grind through a ¼-inch/6-millimeter (medium) die into the bowl of a stand mixer. Using the paddle attachment, mix on medium speed for about 2 minutes, adding the wine as you do so.

**4.** Tie one end of each casing, using a bubble knot (see page 122). Stuff the sausage into the casings and tie each one off with another bubble knot, being careful not to stuff the sausage too tightly. Tie each sausage in the middle to create four 9-inch/23-centimer links (this will allow them to bend in the pot). In a suitable vessel, poach them in stock or water at 170 degrees F./75 degrees C. to an internal temperature of 150 degrees F./66 degrees C., about an hour. Serve hot, or chill and reheat to serve.

**Yield: Two 2.5-pound/1120-gram sausages**

# Venison Salami Sticks

Venison is the general term for deer, elk, and moose, all part of the same ruminant-mammal family. The Midwest teems with whitetail deer, and each season, hunters search for this prize. But we've been surprised by how many hunters have no idea what to do with their kill. We believe that part of hunting is the cooking and the eating—otherwise, it is a waste of the animal and a waste of life. "Field to table" is one of the great culinary experiences. This recipe is a variation on the commercial smoked snack-stick found at party stores and gas stations, but a whole lot better tasting and better for you. The sausage is stuffed into sheep casings, half the size of hog casings, which are often used for breakfast links (see Sources, page 267). This recipe calls for cold- and hot-smoking, but if you're only able to hot-smoke, that's fine too.

**3 pounds/1360 grams boneless venison shoulder, all sinew and fat removed, cut into large dice**
**2 pounds/900 grams pork back fat, cut into large dice**
**2 ounces/56 grams sea salt**
**1 teaspoon/7 grams pink salt**
**1 tablespoon/16 grams coarsely ground black pepper**
**4 garlic cloves, minced**
**2 tablespoons/16 grams mustard seeds, toasted and ground**
**1 teaspoon/2 grams freshly grated nutmeg**
**1 teaspoon/2 grams ground cardamom, toasted and ground**
**1 teaspoon/2 grams coriander seeds, toasted and ground**

**20 feet/6 meters sheep casings, soaked in tepid water for at least 20 minutes and rinsed**

*continued on next page*

**1.** Partially freeze the venison and fat.

**2.** Combine venison, fat, salt, pink salt, pepper, garlic, mustard seeds, nutmeg, cardamom, and coriander and toss together to distribute the seasoning. Grind through a ¼-inch/6-millimeter (medium) die into the bowl of a stand mixer. Using the paddle attachment, mix on medium speed until the ingredients are well distributed, 1 to 2 minutes.

**3.** Stuff the sausage mixture into the sheep casings and tie off into 12-inch/30-centimeter lengths.

**4.** Cold-smoke the sausages for 2 hours.

**5.** Hot-smoke the sausages to an internal temperature of 150 degrees F./66 degrees C. Let cool, and refrigerate until thoroughly chilled. Can be kept for up to 10 days.

**Yield: About 5 pounds/2250 grams sausage; about forty 6-inch/15-centimeter links**

# Venison Salami

This sausage is a good way to use the trim from butchering a deer. Most game meat is very lean, so adding fat is essential for maximum flavor and juiciness. This recipe calls for grinding the meat first without the fat, and then a second time with the fat, which results in a distinct speckling of fat in the salami.

**3.5 pounds/1590 grams lean venison, fat and heavy sinew removed, cut into large dice**
**2 ounces/56 grams sea salt**
**1 teaspoon/7 grams pink salt**
**2 teaspoons/12 grams finely ground black pepper**
**1 tablespoon/6 grams freshly grated nutmeg**
**2 large garlic cloves, minced**
**1.5 pounds/680 grams pork back fat, cut into large dice**

**Two 18-inch/45-centimeter lengths beef middle, soaked in tepid water for at least 20 minutes and rinsed**

1. Combine the venison, salt, pink salt, pepper, nutmeg, and garlic and grind through a ¼-inch/6-millimeter (medium) die into the bowl of a stand mixer. Using the paddle attachment, mix for 2 minutes on medium speed until ingredients are well distributed. Cover with plastic wrap, pressing down on the wrap so that it's in contact with the surface of the meat, and refrigerate for 8 to 24 hours.

2. Paddle in the diced fat. Grind the mixture through the same die back into the mixer bowl. Mix on medium speed for 1 minute, or until the ingredients are well distributed.

3. Tie one end of each casing, using a bubble knot (see page 122). Stuff the sausage into the casings and tie each one off with another bubble knot. Put sausage on a baking sheet and refrigerate, uncovered, overnight.

*continued on next page*

**4.** Cold-smoke the sausages for 2 hours.

**5.** Hot-smoke to an internal temperature of 150 degrees F./65 degrees C., then chill completely in an ice bath.

**Yield: Two 1.75-pound/800-gram salami**

# Mazzafegati

Mazzafegati are fresh liver sausages from Umbria, in central Italy. They're delicious grilled or roasted over an open fire, especially a wood fire. This is an easy sausage to make and we love the addition of pine nuts. Liver is very moist, so it helps to make sure that everything is very cold when you grind the meat and stuff the sausages.

**3 pounds/1360 grams pork shoulder butt, cut into large dice**
**1.5 pounds/680 grams pork back fat, cut into large dice**
**1 pound/450 grams pork liver, outer membrane and veins removed, cut into large dice**
**1.5 ounces/42 grams sea salt**
**2 tablespoons/28 grams sugar**
**1 tablespoon/6 grams finely ground black pepper**
**2 tablespoons/15 grams coriander seed, toasted and ground**
**¼ teaspoon/1 gram ground mace**
**3 garlic cloves, minced**
**Grated zest of 3 oranges**
**⅓ cup/56 grams pine nuts, toasted**
**1 cup/250 milliliters sweet white wine, such as Muscat**

**10 feet/3 meters hog casings, soaked in tepid water for at least 20 minutes and rinsed**

**1.** Combine the pork, fat, liver, salt, sugar, pepper, coriander, mace, garlic, and orange zest in a bowl and toss together. Cover and refrigerate overnight.

**2.** Grind the pork mixture through a ⅛-inch/3-millimeter (small) or your smallest die into the bowl of a stand mixer. Using the paddle attachment, mix on medium speed for about a minute, adding the wine and pine nuts as you do so, until all the ingredients are evenly distributed.

*continued on next page*

**3.** Stuff the sausage into the casings immediately and tie off into 6-inch/15-centimeter links.

**4.** Poach the sausages in 170 degree F./77 degree C. water to an internal temperature of 150 F./66 degree C. Chill in an ice bath and refrigerate until ready to cook.

**5.** Grill over a low fire until heated through.

**Yield: About 5 pounds/2250 grams sausage; about twenty 6-inch/15-centimeter links**

# Salami Cotto

*Cotto* means cooked in Italian. In this case, we use the smoker with a light coating of smoke, or smudge. It should have a hint of smoke while cooking slowly so as not to lose its fat, since it has considerably less fat than usual. This is most like salami you'll find in the grocery store.

4 pounds/1815 grams lean beef shoulder or round, cut into large dice
1.5 ounces/42 grams sea salt
1 teaspoon/7 grams pink salt
2 teaspoons/12 grams finely ground black pepper
2 teaspoons/8 grams ground caraway
3 large garlic cloves, minced
1 pound/450 grams pork back fat, cut into large dice
1 cup/225 milliliters ice water

Two 18-inch/45-centimeter lengths beef middle, soaked in tepid water for at least 20 minutes and rinsed

**1.** Combine the beef, salt, pink salt, pepper, caraway, and garlic and grind through a ¼-inch/6-millimeter (medium) die into the bowl of a stand mixer. Using the paddle attachment, mix for 2 minutes on medium speed until the ingredients are well distributed. Cover with plastic wrap, pressing down on the wrap so that it's in contact with the surface of the meat, and refrigerate for 8 to 24 hours.

**2.** Add the fat and mix on medium, adding water as you do until everything is well distributed. Grind the mixture through the same die back into the mixer bowl. Mix on medium for 1 minute, or until the ingredients are well distributed.

**3.** Tie one end of each casing with a bubble knot (see page 122). Stuff the sausage into the casings and tie each one off with another bubble knot. Refrigerate, uncovered, overnight.

*continued on next page*

**4.** Cold-smoke the sausages for 2 hours (see page 75).

**5.** Hot-smoke to an internal temperature of 150 degrees F./65 degrees C. Then chill thoroughly in an ice bath.

**Yield: Two 1.75-pound/800-gram salami**

# 5. Cooking with and Serving Salumi

Whether you have cured it yourself or are fortunate enough to get your hands on some exquisite imported or homegrown lardo, salami, soppressata, prosciutto, pancetta, spalla, guanciale, or coppa—well, what then? Perhaps you and your companions, all of you circling the board like jackals eager to devour your treasure, cannot wait to eat it right then and there. A perfectly understandable response to great salumi. But, given some time to pause and reflect, what are the best ways to put these special goods to use?

The most common way to revel in the glories of salty, dry-cured meat and fat, of course, *is* to eat it unadorned. By itself. Alone. For us, it is akin to a religious offering. Slip a paper-thin slice of salami or soppressata or prosciutto onto your tongue and let the fat melt and the salt and flavors soak into your tongue and then chew and savor as slowly, as deliberately, as thoughtfully, as lingeringly as possible.

Have a piece of bread with it, and a juice glass of Brunello. If you want to get fancy, put out some cornichons, some mustard, one of the accompaniments on pages 208–12, or something acidic to balance the salty fat, but if you've got wonderful salumi, all you really need is bread and wine. When we traveled the various regions—Emilia-Romagna being the spiritual center of the practice of salumi—we invariably tasted salumi unadorned like this. Usually we were standing around a table with a big cutting board and a few different offerings, sausages and whole muscle, such as coppa. The salumière would cut some bread, open a bottle, carefully peel the casing off the salami, and cut several slices. We'd stand, taste, talk, drink. Unalloyed pleasure. And this is how we recommend you eat salumi.

But there are, in fact, all kinds of ways to put salumi into action in your kitchen. The hog is an animal of divine bounty. It gives and gives. Salumi too is diverse in its powers. While the finest culatello or salami finocchiona is indeed best savored solo, where would pasta carbonara be without guanciale or pancetta? Can finely shaved prosciutto be enhanced, elevated with sweetness and acidity? Absolutely.

In this final section, we discuss a variety of ways to serve and use salumi, beginning with its most common presentation, on the cutting board, and moving forward to differing pairings and, finally, cooking with salumi.

# Tagliere di Salumi

The *tagliere di salumi*, salumi board, is a grand presentation of a variety of cuts and accompaniments. There are many ways to create an impressive board. Following are a few practical guidelines.

As a general rule, plan on 1 to 2 ounces/30 to 60 grams of sliced cured meat per person as an appetizer before a main course, or double the amount for more substantial hors d'oeuvres.

Some purveyors may slice the salumi for you and that is acceptable, especially if thinness is important (with prosciutto, for instance) and you don't have your own Berkel slicer, the Ferrari of deli meat slicers. But salumi is really best sliced just before it is eaten. If it's sliced too early, it can dry out, and the flavor of the fat can be affected by light, so generally we recommend slicing your own salami and smaller whole muscles such as coppa.

How thick should you slice it? Personal preferences vary, but make no mistake: the thickness of the cut affects the taste and the eating experience of a salami or coppa or prosciutto. Generally, the larger the diameter of the sausage or cut, the thinner the slice should be. Smaller dry-cured sausages can be sliced thicker. Slice sausage that has been stuffed into smaller casings, hog casings, approximately ⅛ inch/3 millimeters thick. It will be necessary to chew these slightly thicker slices a bit longer, allowing you to enjoy them more. Sausages stuffed into larger casings should be sliced thinner, approximately 1⁄16 inch/1.5 millimeters, so they are easy to roll up and eat. All dried solid muscles such as coppa and prosciutto should be sliced as thin as possible.

The best way to slice salumi is with an electric deli-style slicer. If this is not an option, use a very sharp thin-bladed slicing knife. Never use a serrated knife. Take your time. Try to keep the slices as uniform as possible. If you are slicing meat (prosciutto, coppa) in advance, layer each slice between wax or parchment paper, or at least feather the slices so that they are easily separated; don't stack them, or they'll stick together. Again, this is less important with sausage.

The casing: to peel or not to peel? For small-diameter sausages, it may not be necessary to remove the casing. Use your culinary common sense—it's not usually an issue of safety but rather one of the pleasure of eating. You may not want to eat a strip of moldy casing, even if it is a good mold. If it's a large sausage, it may have been stuffed into a beef casing; beef casings, which we call for in most of our salami recipes, are difficult to chew and should be removed.

To remove the casing, cut off the tip of the sausage, make a slit down the side with a sharp knife—just as much as you want to slice—and peel it away. If there is white mold on the sausage, brush it off before peeling the sausage. If the casing is difficult to peel, wrap the sausage in a damp towel for five to ten minutes—this will rehydrate the casing and make it easy to peel.

Always slice salumi and handle the slices when they are cold—it's much easier to slice and to arrange boards or platters when the fat is a little more solid and is not sticky. But let the platters of meat come to room temperature before serving to bring out the aroma and complex flavors of the cured meat (thinly sliced meat warms up quickly—don't slice so early that the slices dry out).

Arrange the various meats in groups so your guests can see the differences between the various offerings, the orange-fennel lonza here, the spicy coppa there, the coarser-ground salami in between, and so on.

And, for a great presentation and eating experience, prepare some of the following accompaniments, especially suited to dry-cured meats.

# Olive Oil–Poached Artichokes

These are delicious just on their own, but they can be used in a salad as well. Hard to go wrong with these babies. They work marvelously on a salumi board. Artichokes, exquisitely flavored, are a bit of work, but they are worth it. The poaching oil will be flavored by the artichokes and should be reserved and used for dipping, for vinaigrette or mayonnaise, or for enriching just about anything.

**1 lemon, halved**
**4 large artichokes (about 2 pounds/1000 grams)**
**About 3 cups/750 milliliters extra virgin olive oil**
**Salt and freshly ground black pepper**

**1.** Juice the lemon and add half the juice and the lemon rinds to a large bowl of cold water. To prepare the artichokes, cut off the stem of each one. Snap off a layer or two of outer leaves to make cutting through the artichoke easier, and so that you can gauge where the heart begins. With a sharp knife, cut through the equator of the artichoke just above the heart. Carve out the thistly choke. With a paring knife, starting from the top of the heart and working your way around it, carve off all the tough green leaves so that you're left with just the heart. Artichokes oxidize quickly; as you finish each one, put it in the lemon water to keep it from darkening.

**2.** Drain and quarter the hearts and put them in a nonreactive saucepan just large enough to hold them. Cover them with olive oil and add the remaining lemon juice. Bring to a simmer over high heat, then reduce the heat to low and cook until the hearts are tender, 40 minutes or so.

**3.** Spoon the cooked chokes from the oil into a container. Pour just enough of the poaching oil over them to cover them well; reserve the remaining oil for another use. Season the artichokes with salt and pepper to taste. Refrigerate for up to 2 weeks.

**Serves 6**

## Shaved Raw Artichokes

Artichokes are found in many varieties all over Italy. Some are eaten raw when picked young. Those varieties are difficult to come by here, but the distinctive taste of raw artichokes can be enjoyed with our larger globe artichokes by thinly shaving the hearts and serving them with a lemon vinaigrette. There's a reason we eat these difficult-to-prepare thistles—they're exquisite.

**2 lemons, halved**
**2 large artichokes (about 1 pound/450 grams)**
**1 cup/225 milliliters extra virgin olive oil**
**Salt and freshly ground black pepper**

**1.** Juice the lemons and combine half the juice with 2 cups/500 milliliters water and the rinds in a small bowl. Clean the artichokes as described in the previous recipe, adding them to the lemon water as you finish each one.

**2.** Put the olive oil in a medium bowl and add the remaining lemon juice. Rest a Japanese mandoline on top of bowl. Cut the trimmed hearts crosswise in half. Carefully shave the artichoke hearts, starting with the flat cut side, into the olive oil, making the slices as thin as possible. (You can slice them with a sharp knife or vegetable peeler if you don't have a mandoline.) Toss the artichokes after you slice each half to ensure the slices are well coated with the lemon juice and olive oil, to keep them from turning brown. Season with salt and pepper to taste.

3. Serve the salad within half an hour.

**Serves 6**

# Marinated Roasted Red Peppers

These roasted peppers are good with just about anything—in salads, on pizza, with cooked sausage, in rice pilaf or couscous, and, of course, with salumi. We also use them pureed in polenta (see page 242).

**2 large red bell peppers**
**¼ cup/60 milliliters extra virgin olive oil**
**1 tablespoon/15 milliliters balsamic vinegar**
**Salt and freshly ground black pepper**

**1.** Roast the peppers over a gas flame, turning frequently, until the skin blisters and turns black on all sides. (If you don't have a gas stove, halve the peppers and broil them skin side up until they're black.) Put in a bowl, cover tightly with plastic wrap, and let steam and cool.

**2.** Remove all the blackened skin from the peppers by rubbing them in paper towels. Cut the peppers in half, remove seeds, and the cores and cut them into thin strips.

**3.** Put the peppers in a bowl and toss with the olive oil, balsamic, and salt and pepper to taste. The peppers can be stored in the refrigerator for up to 1 week.

**Serves 6**

# Caramelized Cipollini with Pine Nuts

Cipollini are small sweet flat onions that grow all over Italy. Although they can be difficult to find here, they're often available at farmers' markets or gourmet shops. But, any small onions will work here, though you may have to adjust the amount of sugar based on their flavor. Try to get onions that are no larger than 1 inch/2.5 centimeters in diameter.

**1 pound/450 grams cipollini onions (not peeled)**
**6 tablespoons/100 milliliters extra virgin olive oil**
**1 tablespoon/10 grams sugar**
**2 tablespoons/30 milliliters red wine vinegar**
**¼ cup/46 grams pine nuts, toasted**

**1.** Trim the root ends of the onion and score a small X in each root end to help prevent the end from bulging out during cooking.

**2.** Blanch the onions in a pot of boiling water for 4 to 5 minutes (blanching the onions makes them easier to peel). Drain and allow to cool, then pinch one end of each onion to remove the skin.

**3.** Pour the olive oil into a large sauté pan set over medium heat. Add the onions and cook, shaking the pan every few minutes, until they begin to caramelize 15 to 20 minutes. Add the sugar and stir until it dissolves, then add the vinegar. Cover the pan, turn the heat to low, and cook the onions until they are tender, about 30 minutes. Let them cool in the pan,

**4.** Stir in the pine nuts. The onions can be refrigerated for up to a week. Allow them to come to room temperature before serving.

**Serves 6**

## Belgian Endive Marmalade

Marmalade by definition usually refers to a fruit, mainly citrus, preserve. Here I take sliced Belgian endive and cook it down with sugar, raisins, and vinegar. The bitter taste of the endive marries well with the sweet raisins and sour vinegar. Serve chilled with any dried meat or salami.

**1 tablespoon/16 milliliters extra virgin olive oil**
**18 ounces/500 grams Belgian endive, cut into slices ¼ inch/6 millimeters wide**
**2 ounces/56 grams brown sugar, light**
**3 ounces/84 grams golden raisins**
**¼ cup/65 milliliters champagne vinegar**
**¼ cup/65 milliliters cider vinegar**
**Salt and pepper to taste**

**1.** In a heavy bottomed skillet, heat the olive oil over medium heat. Add the endive and sauté for 5 minutes until softened.

**2.** Add the sugar and sauté 3 more minutes, stirring constantly, being careful not to burn the mixture.

**3.** Add the remaining ingredients and bring to a boil. Turn the heat down and simmer gently until all the liquid has evaporated and the mixture has the consistency of jam. Season with salt and pepper. Chill.

**4.** The marmalade will keep covered in the refrigerator for up to 1 month.

**Yield: 1 cup/225 milliliters**

# Crostini

*Crostini* translates as little toasts or croutons. They're thin slices of bread that are grilled or toasted and used as a vehicle for other ingredients. As with bruschetta, there's very little savory food that doesn't go well on crostini. Here we use ingredients that pair well with salumi. Typically crostini are served as a canapé, but they can also garnish soup or a stew.

## Crostini

**1 baguette, preferably one from a local artisan baker**

**1.** Preheat the oven to 250 degrees F./120 degrees C.

**2.** Using a serrated knife, cut the baguette into as many slices as you need, whatever thickness you wish. We prefer crostini that are about ¼ inch/6 millimeters thick.

**3.** Lay the slices on a rack set on a baking sheet (or directly on the pan if you don't have a rack) and bake until completely dry and crisp, about 20 minutes.

# Roasted Garlic

**1 head garlic**
**1 teaspoon extra virgin olive oil**

**1.** Preheat the oven to 350 degrees F./180 degrees C.

**2.** Rub the garlic with the olive oil and wrap tightly in foil. Roast for 30 to 40 minutes, until the garlic is tender.

3. To use the garlic, simply squeeze it out of its skin. Roasted garlic will keep, tightly covered, in the refrigerator for up to a week.

# Salami with White Bean and Roasted Garlic Crostini

This simple white bean puree flavored with roasted garlic can be made well in advance. Here it is an enrichment for the salami, but it might just as easily be a dip, or one of several spreads to serve on crostini by themselves as a canapé.

**1 cup (7 ounces)/200 grams dried great northern or other white beans, picked over and rinsed**
**1 head roasted garlic (page 214)**
**¼ cup/60 milliliters extra virgin olive oil**
**Salt and freshly ground black pepper**
**12 crostini (page 213)**
**12 thin slices salami (about 2 ounces/56 grams)**

**1.** Preheat the oven to 325 degrees F./160 degrees C.

**2.** Put the beans in a pot, cover with plenty of water, and bring to a boil. Reduce the heat and simmer for 1½ hours, or until the beans are very soft.

**3.** Drain the beans, reserving about 1 cup/250 milliliters of the cooking liquid. Put the beans in a food processor and season with salt and pepper. Cut the roasted garlic head crosswise in half and squeeze the cloves into the food processor. With the machine running, add the olive oil and enough of the cooking liquid until you have a smooth puree with a spreadable consistency. The puree will keep for a week refrigerated.

**4.** Spread some of the bean puree on each crostini and top with the salami. Extra or leftover bean puree can be used as a dip or added as a flavorful thickener to soups and stews.

**Serves 6**

## Crostini of Salami with Tapenade and Roasted Peppers

Roasted peppers have a sweetness that complements the saltiness of the salami; it also balances the salty richness of the tapenade. Ribbons of fresh basil add color and flavor.

**Tapenade as needed (recipe follows)**
**12 crostini (page 213)**
**1 cup/200 grams sliced Marinated Roasted Red Peppers (page 210)**
**12 thin slices salami (about 2 ounces/56 grams) cut into thin strips**
**6 large chiffonade (slivers) fresh basil leaves**

**1.** Spread the tapenade evenly over the crostini and top with the roasted peppers.

**2.** Toss the salami with the basil and arrange on top of the red peppers.

Serves 6

### Tapenade

**1 cup/110 grams pitted Taggiasca olives or other good brine-cured black olives**
**1 anchovy, preferably white**
**1 clove roasted garlic (page 214)**
**1 tablespoon/14 grams drained capers**
**2 fresh basil leaves**
**1 tablespoon/15 milliliters freshly squeezed lemon juice**
**3 tablespoons/45 milliliters extra virgin olive oil**

Combine all the ingredients in a food processor and puree until smooth.

**Yield: 1 cup/225 grams**

## Crostini with Coppa, Tomato, Mozzarella, and Pesto

Here the elements of a Caprese salad serve as the counterpoint to thinly sliced coppa. Fresh mozzarella slices tossed with pesto are arranged on the crostini, topped with tomato slices, and finished with the coppa. If you have access to genuine buffalo mozzarella, use it!

The pesto recipe makes more than you'll need for the crostini, but we figure it's worth making a little extra while you're at it to toss with pasta (or freeze it for up to 3 months, until you're ready to use it). Use a fresh, high-quality, fruity extra virgin oil for the pesto.

**Twelve ¼-inch/6-millimeter-thick slices fresh mozzarella, preferably buffalo mozzarella (about 4 ounces/140 grams)**
**¾ cup/260 grams pesto (recipe follows)**
**Extra virgin olive oil if needed**
**12 crostini (page 213)**
**1 large Roma (plum) tomato, cut into 12 even slices and lightly salted**
**12 thin slices coppa (about 2 ounces/56 grams)**

**1.** Put the mozzarella slices in a bowl and toss with the pesto, coating them evenly (you can add a little olive oil if necessary).

**2.** Place a mozzarella slice on each crostini and top with a slice of tomato followed by a slice of coppa.

**Serves 6**

*continued on next page*

## Pesto

1 cup/28 grams fresh basil leaves
2 ounces/48 grams freshly grated Parmigiano-Reggiano (about ½ cup)
½ cup/120 milliliters extra virgin olive oil
⅓ cup/36 grams pine nuts, toasted
3 garlic cloves
Salt and freshly ground black pepper to taste

Combine all the ingredients in a blender and puree until smooth.

**Yield: 1½ cups/350 grams**

## Crostini with Speck and Parmigiano-Reggiano

Speck is dry-cured ham that's been smoked (see our recipe on page 172). Thinly shaved, it goes beautifully with Parmigiano-Reggiano, like prosciutto—but the smoke gives this traditional combination additional depth. Cut the Parmigiano-Reggiano with a vegetable peeler to make enticing curled shavings.

**12 crostini (page 213)**
**Extra virgin olive oil for brushing**
**12 thin slices speck (about 2 ounces/56 grams)**
**12 shavings or thin slices Parmigiano-Reggiano (about 2 ounces/56 grams)**
**2 tablespoons/2 grams chopped fresh chives**

Brush the crostini with olive oil and top each one with a slice of speck. Top the speck with the Parmigiano-Reggiano and sprinkle with the chives.

**Serves 6**

# Crostini con Spuma

*Spuma* translates literally as foam, but this is not the kind of foam you're likely to find in modernist cuisine. A better term for it would be "whipped." We first tasted this at the wonderful Salumeria Falaschi in the Tuscan town of San Miniato, midway between Pisa and Florence. It was called *spuma di gota,* which literally means foam of the jowl, whipped guanciale (cured hog jowl). It was fabulous, and Brian repaired to his kitchen to create a number of similar creations. All the *spumas* here are more akin to a beautiful sausage mousse, one that's both rich and light and spreads smoothly on toasted country bread. He brazenly calls one of his preparations pig butter—fabulous! For cooking, we'd use plain *spuma di gota* (page 221) or *spuma di lardo* (page 223). The jowl has much more fat on it than pancetta, but fatty pancetta can be substituted here as well. If you refrigerate your *spuma*, be sure to allow it to come to room temperature before serving on crostini.

In addition to being spread on crostini as we do here, *spuma* can be used as an enrichment for risotto or polenta. It also makes a splendid cooking medium. Cook eggs in it, or add minced garlic and red peppers to melted spuma and toss with cooked pasta.

**2 heads roasted garlic (page 214)**
**12 crostini (page 213)**
**½ cup/90 grams Spuma di Gota (recipe follows)**
**2 tablespoons/2 grams chopped fresh chives**

**1.** Remove the roasted garlic cloves from the skin and mash into a puree. Spread a thin layer on each crostini.

**2.** Spoon a dollop of *spuma* onto each crostini, and sprinkle with the chopped chives.

**Serves 6**

# Spuma di Gota

Guanciale is sliced and eaten as salumi or, perhaps more commonly, cut like bacon and cooked to flavor other dishes. But its copious fat makes another preparation possible. Ground, mashed, and whipped until it is soft, airy, and creamy, it can be spread on crostini. We've added ricotta to this spuma to lighten it a little and give it a richer texture.

**8 ounces/225 grams guanciale, cubed**
**¼ cup/56 grams ricotta cheese**

**1.** Grind the guanciale through a ⅛-inch/3-millimeter (small) die.

**2.** Puree the guanciale in a food processor for 30 seconds. Add the ricotta cheese and process for another 30 seconds, or until very smooth.

**Yield: 10 ounces/280 grams**

## Soft Spuma (Whipped Pig Butter)

This *spuma* is made from rendered lard, seasoned with onion and salt, so it's very soft and creamy. It is delicious on grilled bread or dolloped onto a grilled pork chop, where it will melt and mingle with the juices. It also makes a fabulous cooking medium. Stored in a tightly sealed container in the fridge, it should keep for 1 to 2 months (the issue will be its picking up off flavors over time rather than any safety concerns).

**2 cups/360 grams pure lard**
**¼ cup/38 grams finely diced sweet onion**
**2 teaspoons/10 grams sea salt**

**1.** Warm the lard just to liquefy it. Put a couple tablespoons of the lard in a small skillet set over medium heat, add the onion, and sauté until soft but not browned, a minute or two. Set aside to cool.

**2.** Meanwhile, put the remaining lard in a baking dish (so that the lard is about ½ inch/1.25 centimeters deep) in the refrigerator and chill until it starts to solidify, 10 to 15 minutes.

**3.** When the lard is firm but not hard (it should be the consistency of room-temperature butter; ideally, thoroughly chill it, leave it on the counter until it's spreadable, and then proceed), transfer it to the bowl of a stand mixer fitted with the whisk and whip on high speed until light and fluffy (like buttercream icing), 6 to 8 minutes. Fold in the onion and salt.

**Yield: 2 ¼ cups/about 400 grams**

# Spuma di Lardo

Lardo made from Mangalitsa back fat works best here, but it can be made with any cured back fat. Mangalitsa, also called the Wooly Pig because of its copious fur, was originally bred in Hungary but is now being raised in America (see page 27). It's prized for the quality and quantity of its fat. If it's been well cured, no additional salt should be necessary, but as always, taste and season accordingly.

**8 ounces/225 grams lardo**
**Sea salt if needed**

**1.** Grind the lardo through a ⅛-inch/3-millimeter (small) die.

**2.** Puree the ground fat in a food processor for 30 seconds. Taste for seasoning, add salt if necessary, and puree for another minute, or until very smooth.

**Yield: 1¼ cups/225 grams**

# Spuma di Mortadella

We ate this *spuma di mortadella* for the first time in a jazz club near the university in Bologna, the oldest extant university in the world, in a city famed for its mortadella. Use the best mortadella you can find. Better yet, make your own (see page 188), and use leftover or end pieces to make *spuma*.

For an interesting appetizer plate, serve mortadella three ways: sliced thin, in chunks lightly browned in a pan, and as *spuma*. Spread the *spuma* on toasted thinly sliced country bread and drizzle with aged balsamic vinegar.

**8 ounces/225 grams mortadella, cut into medium dice, at room temperature**
**2 tablespoons/30 grams ricotta cheese**
**½ teaspoon/1 gram freshly grated nutmeg**
**6 tablespoons/85 milliliters heavy cream**

**1.** Puree the mortadella in a food processor until smooth. (It may bunch up around the blade; if it does, stop processing and scrape it off.) Add the ricotta and nutmeg and continue to process, slowly adding the cream, until you have a rich, light puree.

**2.** Transfer the *spuma* to a bowl and cover it with plastic wrap, pushing the plastic down onto the surface. Refrigerate for at least an hour, or up to 3 days, before serving.

**Yield: 1½ cups/360 grams**

# Pizza

Making pizza at home is easy, delicious fun, and topping it with salumi is a great strategy. Because salumi is dried, the flavors are concentrated, and so, as a topping, salumi is a powerful player. Add heat to the equation, and the flavor only increases in intensity. Pepperoni, of course, is the default "salami" to use on pizza. We use quotation marks because there isn't a salami called pepperoni in Italy; it's likely an adaptation of spicy salami created by Italian immigrants in America. And, like so many of the great foods immigrants brought to America, it got homogenized and "mediocritized" by mass production. Replace pepperoni with an excellent salami, and you will appreciate why pepperoni pizza came into being in the first place.

Because it has a high fat content, fat that renders in heat, salumi enriches pizza in ways that it can't when used on, say, crostini. The white pizza here, for instance, is topped with lardo, pure cured fatback, which becomes almost a finishing fat for the pizza.

It makes sense that a combination of salumi—coppa, salami, and prosciutto—would be great on a pizza. You're simply taking what is already combined on a *tagliere* (salumi board)—bread, cheese, salumi—and adding heat to meld them together. In a way, you're making an edible warm plate for your dry-cured meats. They're not cooked on the pizza—they're added after it's cooked, so that the heat and steam warm the meat and begin to melt the fat. This is one of the best ways to eat salumi we know.

Our pizza dough (page 232) is very easy, very forgiving; it's basically a kind of flatbread that doesn't require a second rise. And it freezes well, so that you could make a double batch of dough, divide it into individual portions, and wrap and freeze the extra dough until you need it (the dough thaws in a couple of hours).

The best option for cooking pizza is a wood-burning oven with a stone hearth, which Brian uses and Michael is jealous of, but most people don't have such a luxury. The next best thing is a pizza stone in your oven, and this does give pizza a tasty distinctive crust. But you can still make pizza on a baking sheet (in fact, the following pizzas were all tested this way). So there's no excuse not to make pizza!

These pizza recipes are for small individual pizzas about 8 inches/20 centimeters in diameter, 1 or 2 servings. Double, triple, or quadruple the quantities of the toppings to make more pizzas.

## Meatza Pizza

This pizza is really a vehicle for serving salumi. Choose a good variety, both whole-muscle cuts and salami. With basil and roasted red peppers, it's festive and satisfying.

**4 ounces/110 grams Basic Pizza Dough (page 232)**
**1 tablespoon/16 milliliters extra virgin olive oil**
**5 slices Roma (plum) tomato (½ tomato)**
**Sea salt**
**½ cup/70 grams buffalo mozzarella, sliced into 6 pieces**
**5 fresh basil leaves**
**¼ cup/25 grams sliced Marinated Roasted Red Peppers (page 210)**
**1 ounce/28 grams Parmigiano-Reggiano cheese, grated (about ¼ cup)**
**2 ounces/56 grams assorted salumi (coppa, prosciutto, etc.), thinly sliced**

**1.** Preheat the oven to 475 degrees F./230 degrees C. Put a pizza stone, preferably, or a baking sheet or pizza pan in the oven to preheat as well.

**2.** Roll the dough out to an even 8-inch/20-centimeter circle. Brush the dough with olive oil. If you don't have a peel, put the dough on a sheet of parchment paper to make it easy to transfer the pizza to the oven.

**3.** Lay the tomato slices on the dough in a spoke pattern and sprinkle them with sea salt. Arrange the mozzarella, basil, and peppers over the tomato slices, and sprinkle the Parmesan evenly over the top.

**4.** Bake for 8 to 12 minutes, until the dough is nicely browned and the cheese is melted.

**5.** Lay the thinly sliced meats on top, allowing the heat from the pizza to soften them. Cut into 4 to 8 pieces and serve.

**Yield: One 8-inch/20-centimeter pizza, enough for 1 personal pizza or 2 appetizer servings**

# Prosciutto-Pesto Pizza

Prosciutto and pesto are a fine combination, here enhanced with the acidity of the tomatoes and the peppery arugula.

**4 ounces/110 grams Basic Pizza Dough (page 232)**
**2 tablespoons/25 grams Pesto (page 218)**
**5 slices Roma (plum) tomato (½ tomato)**
**Sea salt**
**1 ounce/28 grams Parmigiano-Reggiano cheese, grated (about ¼ cup)**
**2 ounces/56 grams thinly sliced prosciutto**
**½ cup/15 grams arugula, cleaned**
**1 tablespoon/16 milliliters extra virgin olive oil**

**1.** Preheat the oven to 475 degrees F./230 degrees C. Put a pizza stone, preferably, or a baking sheet or pizza pan in the oven to preheat as well.

**2.** Roll the dough out to an even 8-inch/20-centimeter circle. If you don't have a peel, put the dough on a sheet of parchment paper to make it easy to transfer the pizza to the oven.

**3.** Brush the dough with the pesto. Lay the tomato slices on the dough in a spoke pattern and sprinkle them with sea salt. Sprinkle the cheese evenly on top.

**4.** Bake for 8 to 12 minutes, until the dough is nicely browned and the cheese is melted.

**5.** Remove from the oven and lay the prosciutto on top, allowing the heat from the pizza to soften it. Cut the pizza into 8 pieces.

**6.** Toss the arugula with a pinch of salt and the olive oil and arrange on top of the pizza. Serve.

**Yield: One 8-inch/20-centimeter pizza, enough for 1 personal pizza or 2 appetizer servings**

# White Pizza

The simplest of all of these pizzas and probably the most delicious—make it only if you have excellent lardo.

**4 ounces/110 grams Basic Pizza Dough (page 232)**
**1 tablespoon/16 milliliters extra virgin olive oil**
**Sea salt**
**2.5 ounces/70 grams buffalo mozzarella, sliced into 6 pieces**
**1 ounce/28 grams Parmigiano-Reggiano cheese, grated (about ¼ cup)**
**1.5 ounce/42 grams lardo, thinly sliced**

**1.** Preheat the oven to 475 degrees F./230 degrees C. Put a pizza stone, preferably, or a baking sheet or pizza pan in the oven to preheat as well.

**2.** Roll the dough out to an even 8-inch/20-centimeter circle. Brush the dough with the olive oil. If you don't have a peel, put the dough on a sheet of parchment paper to make it easy to transfer the pizza to the oven.

**3.** Sprinkle the dough with sea salt, and lay the mozzarella slices evenly on the dough. Sprinkle with the Parmesan.

**4.** Bake for 8 to 12 minutes, until the dough is nicely browned and the cheese is melted.

**5.** Remove from the oven and lay the lardo on top, allowing the heat from the pizza to soften and melt it slightly. Cut into 8 pieces and serve.

**Yield: One 8-inch/20-centimeter pizza, enough for 1 personal pizza or 2 appetizer servings**

# Speck and Vegetable Pizza

Speck, smoked prosciutto, makes an especially good pizza topping. Here it is combined with roasted mushrooms and peppers, basil, and mozzarella.

**3 tablespoons/45 milliliters canola oil**
**8 ounces/240 grams button mushrooms**
**Salt and pepper to taste**
**4 ounces/110 grams Basic Pizza Dough (page 232)**
**1 tablespoon/16 milliliters extra virgin olive oil**
**5 slices Roma (plum) tomato (½ tomato)**
**Sea salt**
**5 fresh basil leaves**
**¼ cup/28 grams sliced Marinated Roasted Red Peppers (page 210)**
**½ cup/70 grams whole-milk mozzarella, shredded**
**2 ounces/56 grams speck (or other air-dried meat, such as lonza or prosciutto), thinly sliced**

**1.** Heat a large dry stainless steel skillet over high heat until it's very hot, about 5 minutes. Add the canola oil (it should immediately be on the verge of smoking). Add the mushrooms in one layer, pressing down on them to sear them well. When they have a good sear, flip them and continue cooking until seared on both sides. Season with salt and plenty of pepper, transfer to a plate, and allow to cool. (The mushrooms will keep, tightly covered, in the refrigerator for a week.)

**2.** Preheat the oven to 475 degrees F./230 degrees C. Put a pizza stone, preferably, or a baking sheet or pizza pan in the oven to preheat as well.

**3.** Roll the dough out to an even 8-inch/20-centimeter circle. Brush the dough with the olive oil. If you don't have a peel, put the dough on a sheet of parchment paper to make it easy to transfer the pizza to the oven.

**4.** Lay the tomato slices on the dough in a spoke pattern and sprinkle with sea salt. Arrange the basil, mushrooms, and peppers over the tomato slices. Arrange the mozzarella evenly over the top.

**5.** Bake for 8 to 12 minutes, until the dough is nicely browned and the cheese is melted.

**6.** Remove from the oven and lay the speck on top, allowing the heat from the pizza to soften the meat. Cut into 8 pieces and serve.

**Yield: One 8-inch/20-centimeter pizza, enough for 1 personal pizza or 2 appetizer servings**

## Basic Pizza Dough

This recipe is adapted from Michael's *Ratio: The Simple Codes Behind the Craft of Everyday Cooking*. It is indeed, simple: 5 parts flour, 3 parts water (it's warm to hasten the yeast's activity, though the temperature isn't critical), yeast, salt, a little honey for sweetness, and some olive oil for a flavorful enrichment.

There are only a few fundamentals involved in making great pizza dough: Measure the ingredients by weight for consistency. Mix the dough in a stand mixer or by hand until it is so elastic you can stretch it to translucency. Let it rise until it's doubled in size. And that's it, you're ready to go. If you want to make the dough a day ahead, you can refrigerate it overnight, and it will only develop more complex flavors.

**1.25 pounds/500 grams bread flour**
**2 teaspoons/14 grams fine sea salt**
**12 ounces/375 milliliters warm water**
**1 teaspoon/6 grams active dry yeast**
**1 tablespoon/21 grams honey**
**2 tablespoons/30 milliliters extra virgin olive oil**

**1.** Combine the flour and salt in the bowl of a stand mixer fitted with the dough hook.

**2.** Combine the warm water, yeast, and honey in small bowl and stir to dissolve the yeast. Add to the flour, add the olive oil, and mix on medium speed for 10 minutes, or until the dough is smooth and elastic. If you cut off a section of the dough you should be able to stretch it until it becomes translucent before it tears.

**3.** Remove the bowl from the mixer, cover with plastic wrap or a moist towel, and let rise in a warm place until the dough doubles in size, 1 to 2 hours, depending on how warm it is.

**4.** Punch the dough down to release the trapped gas and divide into eight 4-ounce/110-gram balls. Round them on the countertop, pushing them back and forth between cupped hands.

**5.** Spray a baking sheet with vegetable oil spray or line with parchment and arrange the balls of dough on it, leaving space between them. Allow to rest before rolling out the dough. If you are not going to use the dough right away, cover and refrigerate overnight or wrap individually and freeze; allow the dough to come to room temperature before rolling it out.

**Yield: 2 pounds/800 grams dough; enough for eight 4-ounce/110-gram pizzas**

# Pasta and Polenta

Salumi are a natural pairing with starches, which carry their flavor well. Salumi with pasta couldn't be easier. Salumi with beans, for that matter, with potatoes, with corn, even with robust vegetables like cabbage and cauliflower—salumi makes everything taste better!

Even that statement is one word too long. How about this: salumi makes everything better.

## Classic Spaghetti Carbonara

Carbonara is one of our all-time favorite pasta dishes. It's simple to prepare, economical, and deeply satisfying, because of the salumi used to flavor it. We like the classic guanciale, cured hog jowl, as is customary in Rome, but it's excellent if you want to use pancetta or even smoked bacon instead. Pasta, rendered fat, and crisp chewy meat from the jowl or the belly, eggs, and cheese—that's classic carbonara. Brian likes some garlic in his, and Michael loves the freshness and color that parsley gives it. The eggs are cooked by the heat of the pasta, but because they're very loosely cooked, we recommend farm or organic eggs for this.

**1 pound/450 grams spaghetti**
**2 tablespoons/32 milliliters extra virgin olive oil**
**4 ounces/110 grams guanciale, cut into strips**
**4 large garlic cloves, minced**
**2 large eggs**
**4 ounces/110 grams Parmigiano-Reggiano cheese, grated (about 1 cup)**
**Freshly ground black pepper**
**Chopped fresh parsley for sprinkling**

**1.** Cook the spaghetti in a large pot of boiling salted water until just al dente. Drain, reserving ¼ to ½ cup/60 to 120 milliliters of the pasta water.

**2.** Meanwhile, heat the oil in a large deep skillet over medium heat until hot. Add the guanciale and sauté until crisp, 8 to 10 minutes. Add the garlic and cook just until softened.

**3.** Add the spaghetti to the skillet, along with the reserved pasta water, just enough to keep it moist. Toss with the guanciale, and cook for 2 minutes.

**4.** Meanwhile, beat the eggs with ¾ cup/80 grams of the cheese in a bowl.

**5.** Remove the spaghetti from the heat and pour the eggs over it, mixing quickly, looking for eggs to thicken. (Do not reheat the pan after the eggs are added.) The sauce should be creamy but not terribly thick. Season with pepper to taste, sprinkle with parsley and the remaining ¼ cup/30 grams Parmesan cheese, and serve.

**Serves 4**

## Potato Gnocchi with Pancetta, Gorgonzola, and Pine Nuts

These classic potato gnocchi are served with a rich Gorgonzola sauce and garnished with pine nuts. Like the carbonara on page 234, this pasta gets its depth of flavor from the pancetta.

Gnocchi can be heavy and gluey if not made well. Peel and rice the potatoes while they are still warm, and be sure not to overwork the dough after you've added the flour.

**THE GNOCCHI**

**2 pounds/900 grams russet (baking) potatoes, scrubbed**
**1 large egg**
**1¾ cups/210 grams cake flour**
**Salt and freshly ground black pepper**
**Cayenne pepper**
**Olive oil**

**THE SAUCE**

**4 ounces/110 grams pancetta, diced**
**2 tablespoons/32 milliliters extra virgin olive oil**
**2 cups/480 milliliters heavy cream**
**8 ounces/225 grams Gorgonzola cheese**
**½ cup/110 grams pine nuts, toasted**

**1.** Put the potatoes in a pot with cold water to cover and bring to a boil, then reduce the heat and simmer gently until tender, 30 to 35 minutes. Drain the potatoes and let cool slightly.

**2.** As soon as you are able to handle them, peel the potatoes. Pass them through a food mill or a ricer and transfer to the bowl of a stand mixer fitted with the paddle attachment. Add the egg and mix on the lowest speed just until incorporated. Scrape down the sides of the bowl, then slowly add the flour, mixing just until incorporated. Season to taste with salt, pepper, and cayenne. Do not overmix, or the gnocchi will be heavy. (If you don't have a stand mixer, this can easily be done by hand.)

**3.** On a floured surface, roll the dough into ropes about 1 inch/2.5 centimeters in diameter. Cut into 1-inch/2.5-centimeter pieces and roll each piece over the back of a wicker basket, the back of a fork, or a gnocchi paddle to create grooves.

**4.** Bring a large pot of generously salted water to a boil. Prepare an ice water bath. (Choose a bowl big enough to hold your colander or strainer, so you can submerge the gnocchi in the ice water and then remove them by simply lifting the colander out of the ice bath.)

**5.** Drop the gnocchi into the boiling water and cook until they float, about 5 to 8 minutes. Scoop them out, and chill them in the ice bath. Remove them from the ice bath, drain well, and toss lightly with olive oil to prevent sticking.

**6.** Sauté the pancetta in the olive oil in a large heavy-bottomed skillet over medium heat until crispy, about 5 to 10 minutes. Add the heavy cream and bring to a boil, then crumble in the Gorgonzola. Turn the heat down and cook until the sauce thickens.

**7.** Fold in just enough gnocchi so the sauce evenly coats each dumpling and heat through. Serve garnished with the pine nuts. Refrigerate any leftover gnocchi, covered, for up to 3 days.

**Serves 8**

## Butternut Squash Ravioli with Prosciutto and Sage Brown Butter

The butternut squash filling for this ravioli is sweet from brown sugar and salty-savory from prosciutto. The ravioli is served in a simple brown butter sauce with sage, one of the great all-purpose sauces for pasta and potatoes.

**THE PASTA DOUGH**

**9 ounces (about 1 cup)/250 grams all-purpose flour**

**3 large eggs**

**1 egg, beaten, for egg wash**

**THE FILLING**

**1 butternut squash (3 pounds/1300 grams)**

**¼ cup/65 milliliters extra virgin olive oil**

**2 teaspoons/6 grams dark brown sugar**

**1 large egg**

**¼ cup/56 grams chèvre (goat cheese)**

**2 tablespoons/56 grams ricotta**

**½ cup/28 grams panko crumbs**

**1 teaspoon/12 grams minced shallots**

**Salt and freshly ground pepper**

**Freshly grated nutmeg**

**4 ounces/110 grams prosciutto, cut into small dice**

**THE SAUCE**

**8 tablespoons/112 grams unsalted butter**

**1 tablespoon/4 grams julienned fresh sage leaves**

**2 ounces/60 grams thinly sliced prosciutto, cut into fine julienne, for garnish**

**1.** Preheat the oven to 350 degrees./180 degrees C.

**2.** Cut butternut squash lengthwise in half and scoop out the seeds. Put cut side up in a baking pan, brush with the oil, and sprinkle with the brown sugar. Roast for about 1 hour, until soft. Let cool to room temperature.

**3.** Scoop the flesh out of the skin, and squeeze out excess moisture. Put in a food processor, add the egg, cheeses, panko crumbs, and shallots and process to a puree.

**4.** Transfer the puree to a bowl and stir in salt, pepper, and nutmeg to taste. Fold in the prosciutto. Set aside.

**5.** Put the flour in a bowl and make a well in the middle. Crack the eggs into a cup and pour them into the well. Using your fingers, stir the eggs, gradually incorporating the flour. (Or put the flour in a food processor, add the eggs, and process just to mix.)

**6.** Turn the dough out onto a floured surface and knead until velvety smooth, 5 to 10 minutes. Wrap in plastic and let rest in the refrigerator for 10 minutes to 1 hour.

**7.** Cut the dough into 2 or 3 pieces. Using a pasta machine, roll out the dough into very thin sheets, optimally about 5 inches/12 centimeters wide; cut into sheets 18 inches/45 centimeters long.

**8.** Lay a sheet of pasta over the bottom of a ravioli form. Fill, using about a tablespoonful of filling per ravioli. Lay a second sheet of pasta over the top, and press, forming perfectly square pillows. (If you don't have a ravioli press, you can cut the ravioli with a sharp knife or a biscuit cutter.) Transfer to a baking sheet and repeat with the remaining dough and filling, using more baking sheets if necessary so the ravioli are in one layer. Freeze.

**9.** When ready to cook and serve, bring a large pot of generously salted water to a boil. Drop in the ravioli, being careful not to crowd them, and cook until they float.

**10.** Meanwhile, melt the butter with the sage in a large skillet over medium heat. Reduce the heat and cook, stirring occasionally, until the butter is a rich brown color.

**11.** When the ravioli are done, lift them out and add them to the sage butter sauce; add a couple spoonfuls of the pasta water if needed.

**12.** Arrange the ravioli on a platter or individual plates, garnish with the julienned prosciutto, and serve.

**Serves 4**

## Semolina Crescents with Pancetta Caponata

Semolina is coarse flour made from durum wheat. While it's commonly used to make pasta, it can also be cooked like polenta, then spread in a pan to cool and solidify. You can then cut the finished semolina into any shapes you wish. Here the semolina, enriched with Parmesan and butter, is cut into crescents, topped with more cheese reheated with caponata and with the eggplant discs, then cloaked in speck. This makes a great starter or side dish.

**SEMOLINA**
**1¼ cups/296 milliliters water**
**5 ounces/142 grams semolina flour**
**2 ounces/56 grams Parmigiano-Reggiano cheese, grated (about ½ cup)**
**2 tablespoons/28 grams unsalted butter, softened**
**Salt and freshly ground black pepper**

**1 cup or ¼ recipe Caponata with Pancetta (page 253), coarsely chopped**
**4 ounces/110 grams Parmigiano-Reggiano, grated (about 1 cup)**
**8 large thin slices (about 4 ounces/110 grams) speck**

**1.** Bring the water to a boil in a medium saucepan. Slowly pour in the semolina, stirring constantly, then reduce the heat to medium-low and cook, stirring frequently, for 25 minutes, or until the semolina is cooked (taste is the best way to determine doneness) and creamy.

**2.** Whisk in the cheese and butter, and salt and pepper to taste. Pour into a greased 9-inch/23-centimeter square baking pan. Allow to cool, then chill until set.

**3.** Preheat the oven to 350 degrees F./180 degrees C.

**4.** Using a 3-inch/7.5-centimeter round cutter, cut 8 crescent shapes from the semolina. Arrange on a parchment-lined baking sheet. Spoon some caponata onto each crescent, then sprinkle with the Parmesan and season with salt.

**5.** Bake for 12 to 15 minutes, or until the cheese is melted and the semolina is warm all the way through. Place a slice of speck on top of each crescent, allowing the residual heat to soften the ham, and serve.

**Serves 8**

# Red Pepper Polenta with Salumi

This is a great side dish for most meats and fish. The polenta is brightly colored and beautifully flavored by the roasted red peppers, which balance the intense salty flavor of the salumi.

**2 cups/480 milliliters chicken stock (page 244)**
**2 tablespoons/28 grams unsalted butter**
**1 cup (6 ounces) 165 grams cornmeal**
**¼ cup/65 milliliters Marinated Roasted Red Peppers (page 210)**
**4 ounces/110 grams assorted salumi (such as coppa, prosciutto, lonza, and salami), cut into fine julienne**
**Salt and freshly ground black pepper**
**½ cup/28 grams fresh basil leaves, cut into chiffonade (slivers)**
**1 ounce/28 grams Parmigiano-Reggiano cheese, grated (about ¼ cup)**
**1 ounce/28 grams Gruyère cheese, grated (about ¼ cup)**

**1.** Bring the stock and butter to a boil in a medium saucepan. Slowly pour in the cornmeal, stirring constantly, then reduce the heat to a simmer and cook, stirring constantly, for 35 minutes.

**2.** Puree the red peppers in a food processor. Add them and the salumi to the polenta and cook for 5 minutes. Season with salt and pepper to taste.

**3.** Pour the polenta into a greased 9-inch/23-centimeter square baking pan and let cool, then chill until firm.

**4.** Preheat the oven to 425 degrees F./220 degrees C.

**5.** Cut the polento into 8 squares. Top with the grated cheese. Bake for about 10 minutes, until the cheese is melted and golden brown and the polenta is heated through.

**6.** Garnish with the basil and serve.

**Serves 8**

## Soups and Salads

Soups are among the most satisfying and simple meals to prepare. Given good stock, it's almost impossible to go wrong. It's that "given good stock" provision that trips many people up, but we can't emphasize enough the importance of using good stock and what a difference it makes. And so we've provided an easy chicken stock recipe that works for all these soups. However, we're under no illusion that everyone is going to make stock. If you buy stock, try to find a market in your area that sells fresh or frozen stock. If you must use canned broth, use half canned mixed with half water. If you have time, simmer an onion and two carrots in this mixture for 30 minutes to 1 hour, and then strain.

Actually, soups that include salumi are more tolerant of the canned stuff because the salumi's flavors are so big and add hearty, piquant depth. Almost any soup tastes better with pancetta or guanciale. If you have a prosciutto bone or especially a rind from a bacon or pancetta, add it to soup for extraordinary body. A Parmesan rind will also enhance the flavor of any of the following soups.

Salads are a natural for salumi. Greens dressed with vinaigrette have the acidity to balance the fat and salt. Some lettuces, such as romaine, add crunch to salumi's softness. Other greens, such as arugula and watercress, give a peppery punch. Fruits add a different kind of acidity, plus sweetness.

You can lay thinly sliced salami or coppa or prosciutto on top of just about any salad; or use julienned salumi as a garnish. Or, as in the first recipe here, use sliced salumi as a bed for a salad.

But don't limit yourself to lettuces—legumes, fruit, and cooked vegetables can be part of your salad repertoire as well.

# Chicken Stock

2 pounds chicken bones
2 quarts/2 liters water
1 large onion, chopped
2 or 3 large carrots, chopped
2 stalks celery, chopped
1 bay leaf
1 teaspoon black peppercorns, crushed beneath a sauté pan
2 to 3 garlic cloves
2 to 3 thyme sprigs (optional)
2 to 3 parsley sprigs (optional)

**1.** Put the bones in a large pot (a 6-quart stockpot is good), pour the water over the bones, and bring to a simmer over high heat. Skim the foam and congealed proteins from the surface. Reduce the heat to low and simmer very gently for 4 hours (or place the pot in an oven preheated to below 200 degrees F./93 degrees C.).

**2.** Add the remaining ingredients, bring the stock back up to just below a simmer, and cook for another 45 minutes to an hour.

**3.** Strain the stock, then pass it through a fine-mesh sieve or a sieve lined with cheesecloth. Cool before refrigerating.

**Yield: 1.5 quarts/1.5 liters**

## Vegetable Soup with Acini di Pepe and Pancetta

Acini di pepe is a little bead-like pasta that gives great texture to this soup packed with vegetables. Because it's made with just pasta and vegetables, the power of adding pancetta to this soup—of adding any salumi—is especially evident.

The beans, pasta, and potatoes are all cooked separately and can be prepared up to a day ahead.

**6 ounces/180 grams Yukon Gold potatoes, cut into small dice**
**4 cups/1 liter chicken stock**
**¼ cup/110 grams acini di pepe pasta**
**1 ounce/30 grams pancetta, cut into ¼-inch/6-millimeter dice**
**2 tablespoons/30 milliliters extra virgin olive oil**
**½ cup/60 grams diced (¼-inch/6-millimeter) onion**
**½ cup/60 grams diced (¼-inch/6-millimeter) carrots**
**½ cup/60 grams diced (¼-inch/6-millimeter) celery**
**Salt and freshly ground black pepper**
**1 cup/120 grams tomatoes, peeled, seeded, and diced**
**½ cup/60 grams half-moon slices zucchini**
**½ cup/60 grams green beans, blanched briefly in boiling salted water and cut into 1-inch/2.5-centimeter lengths**
**1 ounce/30 grams any salume, peeled and cut into ¼-inch/6-millimeter dice**
**1 tablespoon/15 milliliters red wine vinegar**

**OPTIONAL GARNISHES**
**About 6 tablespoons pesto (page 218)**
**6 crostini (page 213)**
**Freshly grated Parmigiano-Reggiano cheese**

**1.** Put the potatoes in a saucepan, add enough chicken stock to cover, bring to a boil, and cook until almost tender. Set aside in the broth.

**2.** Cook the pasta in boiling salted water until al dente; drain, rinse, and set aside.

*continued on next page*

**3.** Sauté the pancetta in the olive oil in a heavy-bottomed soup pot until it has rendered its fat. Add the onion and sauté until soft. Add the carrots and celery and cook until tender, 4 to 5 minutes, being careful not to brown the vegetables. Season with salt and pepper.

**4.** Add the tomatoes, zucchini, the remaining chicken stock, the potatoes with their cooking liquid, the pasta, the green beans, the salume, and the red wine vinegar. Bring to a simmer, and simmer for 10 to 15 minutes. Taste and season with more salt if necessary.

**5.** To serve, ladle the soup into bowls. If desired, spread pesto on the crostini, place on top of the soup, and sprinkle with Parmesan.

**Serves 6**

## White Bean and Pasta Soup

Something about white beans makes them especially good in soups—their color, flavor, and texture marry beautifully with the stock. Here they also carry the flavor of the pancetta and fat very well. Once the beans and pasta are cooked—something that can be done up to 3 days ahead—this soup comes together in about 15 minutes.

**1 cup (8 ounces)/225 grams dried cannellini or other white beans, soaked overnight in water to cover**
**8 ounces/225 grams pasta shapes, such as cavatappi or ditalini**
**3 ounces/85 grams thinly sliced pancetta, minced**
**2 ounces/56 grams lardo or pork back fat, minced and mashed to a paste**
**1 cup/110 grams diced (¼-inch/6-millimeter) onion**
**4 garlic cloves, minced**
**¼ cup/28 grams chopped parsley**
**2 tablespoons/40 grams tomato paste**
**8 ounces/225 grams Roma (plum) tomatoes, peeled, seeded, and diced**
**6 cups/1450 milliliters chicken stock**
**Salt and freshly ground black pepper**
**6 tablespoons/96 milliliters extra virgin olive oil**

**1.** Put the beans in a saucepan, cover generously with water, and bring to a boil, then reduce the heat and simmer until tender, about 2 hours. Drain and set aside.

**2.** Meanwhile, cook the pasta in boiling salted water until al dente. Drain, rinse, and set aside.

**3.** Combine the pancetta and lardo in a large skillet and cook over medium-low heat until the pancetta is nicely browned and has rendered its fat. Add the onion and garlic and sauté until golden. Add the parsley and tomato paste and cook for 2 to 3 minutes. Add the tomatoes and chicken stock and bring to a boil, then lower the heat to a simmer. Add the pasta and beans and simmer for 10 minutes. Season with salt and pepper to taste. To serve, ladle the soup into large rimmed bowls and drizzle the olive oil on top.

**Serves 6 to 8**

## Chickpea Soup with Spicy Salami

Chickpeas are, in our opinion, an underused legume. You can use canned chickpeas here, but we recommend you buy dried chickpeas, and cook them yourself. The flavor and texture are worth it. We feature them here with the spicy, paprikay chorizo because, well, chorizo makes everything not just a little better, but a *lot* better. Saffron gives the soup a beautiful yellow color.

**1 cup (8 ounces)/225 grams dried chickpeas, soaked overnight in water to cover**
**½ cup/120 milliliters extra virgin olive oil**
**1 cup/110 grams diced (¼-inch/6-millimeter) onion**
**12 garlic cloves, minced**
**1 tablespoon/4 grams chopped fresh rosemary**
**1 teaspoon/1 gram saffron threads**
**1 cup/170 grams peeled, seeded, and diced small Roma (plum) tomatoes**
**8 ounces/225 grams *salami picante* (or Spanish chorizo) cut into small dice**
**8 cups/2 liters chicken stock**
**Salt and freshly ground black pepper**

**1.** If using dried chickpeas, put them in a saucepan, add water to cover, and bring to a boil, then reduce the heat and simmer until tender, about 3 hours. Drain and set aside.

**2.** In a heavy-bottomed soup pot, heat the olive oil and sauté the onion and garlic until soft, about 10 minutes. Add the rosemary, saffron, and tomatoes and cook for a few minutes, then add the chorizo and cook for another minute or so. Add the chicken stock and chickpeas, bring to a simmer, and cook for 10 minutes. Season with salt and pepper to taste.

**3.** Ladle into soup bowls and serve.

**Serves 8**

## Coppa, Orange, and Onion Salad

A bed of good coppa topped with greens dressed with lemon and olive oil, orange, and onion makes a great starter or a light lunch.

**THE DRESSING**

**1 tablespoon/15 milliliters freshly squeezed lemon juice, or to taste**
**Salt and freshly ground black pepper**
**3 tablespoons/44 milliliters extra virgin olive oil**

**4 ounces/110 grams coppa, thinly sliced**
**16 orange segments, white pith and membrane removed**
**¼ cup/28 grams red onion slivers**
**1 cup/57 grams watercress**
**2 cups/85 grams field greens**

**1.** Combine the lemon juice with salt and pepper to taste in a small bowl and stir to dissolve the salt. Whisk in the olive oil. Taste and adjust with more lemon juice or salt as needed.

**2.** Arrange the coppa on four plates. Combine the orange segments, red onion, watercress, and greens in a bowl and toss with the dressing. Arrange the salad on the coppa and serve.

**Serves 4**

## Spinach with Warm Pancetta Vinaigrette

This is a take on the classic salad of our youth, spinach salad with sliced raw mushrooms and warm bacon vinaigrette. Spinach and pancetta go together in the same way that spinach and bacon go together. The pancetta is heated and lightly browned, then the vinegars and olive oil are combined with the rendered fat to make the dressing. Add chopped hard-boiled egg, and you've got a great bacon-and-egg salad.

**THE DRESSING**

**4 ounces/110 grams thinly sliced pancetta, cut into fine julienne**
**½ cup/120 milliliters extra virgin olive oil**
**2 tablespoons/32 milliliters red wine vinegar**
**1 tablespoon/16 milliliters balsamic vinegar**
**Salt and freshly ground black pepper**

**8 ounces/225 grams button mushrooms, sliced ⅛ inch/3 millimeters thick**
**4 cups (6 ounces)/165 grams baby spinach, cleaned**
**½ cup/56 grams thinly sliced scallions**
**2 hard-boiled eggs, coarsely chopped**

**1.** Cook the pancetta in a large skillet over medium-low heat until it is nicely browned and has rendered its fat. Whisk in the oil and vinegars and remove from the heat. Season with salt and pepper to taste.

**2.** Toss the mushrooms with half the warm dressing in a large bowl. Add the spinach and scallions, then add the remaining dressing and toss to coat.

**3.** Arrange the salad on four plates, top with the chopped hard-boiled eggs, and a sprinkling of black pepper, and serve.

**Serves 4**

# Grilled Radicchio with Pancetta and Balsamic

There's no better lettuce to grill than radicchio. Its small, compact shape holds up well in the heat and here, its bitter notes, combined with the charred flavors of the grill, are balanced by the sweetness of the balsamic vinegar. Pancetta, the base of the warm vinaigrette, brings it all together.

**2 heads radicchio, halved lengthwise**
**1 cup/240 milliliters extra virgin olive oil**
**Salt and freshly ground black pepper**
**4 ounces/110 grams thinly sliced pancetta, diced**
**2 shallots, thinly sliced**
**¼ cup/60 milliliters balsamic vinegar**

1. Heat a grill until hot.

2. Meanwhile, blanch the radicchio in a pot of boiling water for 1 minute. Drain and cool in an ice bath; drain again. Wrap each half in a kitchen towel and squeeze out the excess water.

3. Toss the radicchio with 2 tablespoons/30 milliliters of the olive oil. Season with salt and pepper. Arrange the radicchio cut side down on the grill and grill until lightly browned on each side, about 2 minutes. Arrange the radicchio on a serving plate.

4. Combine the pancetta and 2 tablespoons/30 milliliters olive oil in a medium sauté pan and cook over medium-low heat until the pancetta browns and renders its fat. Add the shallots and sauté just until they wilt. Add the vinegar and simmer until the vinegar is reduced to a glaze. Whisk in the remaining ¾ cup/180 milliliters olive oil and season with salt if necessary (depending on the saltiness of your pancetta, it may not need further seasoning).

5. Pour the warm vinaigrette over the radicchio and serve.

**Serves 4**

## Chickpea, Roasted Garlic, and Guanciale Salad

Another excellent use for the versatile chickpea—a salad with a roasted garlic vinaigrette and guanciale (or pancetta), garnished with fresh basil. It's a great side salad for a board of salumi or plate of sliced sausages.

**1 cup (8 ounces)/225 grams dried chickpeas, soaked and cooked according to the instructions on page 248**
**¼ cup/60 milliliters extra virgin olive oil**
**1 tablespoon/28 grams roasted garlic (page 213)**
**2 Roma (plum) tomatoes, peeled, seeded, and cut into small dice**
**¼ cup/60 grams guanciale or pancetta, sautéed and cooled, fat drained**
**1 tablespoon/16 milliliters balsamic vinegar**
**1 tablespoon/2 grams chiffonade fresh basil leaves**
**Salt and freshly ground black pepper**

**1.** Put the chickpeas in a bowl, add the olive oil and roasted garlic, and mix well. Add the tomatoes, guanciale, balsamic, and basil. Season with salt and pepper and serve.

**Serves 4 (Yield: Approximately 2 cups/450 milliliters)**

## Caponata with Pancetta

Caponata is a mélange of eggplant and other vegetables, typically including celery and tomatoes, with olives and capers. We use onion sautéed with pancetta, and throw in some pine nuts for crunch and flavor.

**1 large eggplant (about 1 pound/450 grams), peeled and cut into 1-inch/2.5-centimeter dice**
**4 ounces/110 grams pancetta, cut into small dice**
**¼ cup/65 milliliters extra virgin olive oil**
**1 cup/225 grams diced (½-inch/1.25-centimeter) onion**
**1 cup/225 grams diced (½-inch/1.25-centimeter) celery**
**1 cup/110 grams diced tomatoes**
**1 cup/110 grams green olives, pitted and chopped**
**1 tablespoon/20 grams drained capers**
**¼ cup/60 milliliters white wine vinegar**
**¼ cup/56 grams pine nuts, toasted**
**¼ cup/8 grams chiffonade fresh basil leaves**
**Salt and freshly ground black pepper**
**Sugar to taste**

**1.** Lightly salt the eggplant and lay on a baking sheet lined with paper towels. Let stand for about 2 hours to remove some of the excess moisture.

**2.** Sauté the pancetta in the olive oil in a large skillet until it is lightly brown and has rendered its fat, 5 to 10 minutes. Add the onion and sauté until just translucent, another 2 minutes. Add the celery and tomatoes and cook for another 4 to 5 minutes. Add the olives, capers, and vinegar and sauté for 1 minute. Toss in the eggplant and sauté for 5 to 8 minutes.

**3.** Fold in the pine nuts and basil and season with salt and pepper and sugar to taste. Serve hot or at room temperature.

**Serves 4**

# Classic Combinations

Some combinations seem so familiar as to not require mentioning, or so ubiquitous that to serve them seems a cliché. But many such pairings, such as melon and prosciutto, are familiar and ubiquitous for a reason. Classic pairings taste great! The following flavor combinations are some of the simple things in life that we can't improve on—and don't want to. Treat each ingredient with respect, and use fruit and vegetables that are at their peak. After all, nature is the true artist and we cooks are the craftsmen who put things together.

## Prosciutto/Coppa with Melon, Figs, Peaches, or Pears

The sweetness of certain fruits and the dense savory saltiness of salumi is a blessed union. Throw in some Parmigiano-Reggiano, and you have a dish. Show off these ingredients to their advantage by serving them together, simply and unadorned. A dish comprising only three ingredients relies fundamentally on the quality of the ingredients. Use good prosciutto or coppa and splurge on real Parmigiano-Reggiano. Pair them only with fruit that's in season and ripe.

The following is a guideline for serving 4; quantities can be scaled up or down as needed. This might be a first course, or set it out as hors d'oeuvres. A common way to serve these items is to cut the fruit into wedges or slices, wrap the prosciutto around each, and arrange on a plate. Or simply suggest that people mix and match as they please.

**Prosciutto or coppa, sliced paper-thin**
**Parmigiano-Reggiano cheese, shaved with a vegetable peeler**
**Ripe melon, figs, peaches, or pears**

We have no specific instructions here other than finding excellent ingredients. In terms of quantities, use common sense. Plan on 2 to 4 ounces/60 to 120 grams each of salumi and fruit per person and 1 to 2 ounces/30 to 60 grams of cheese. Put all the ingredients out on a cutting board, or arrange the ingredients on small plates for an elegant first course.

**Serves 4**

## Coppa with Grapefruit, Arugula, and Olive Oil

The pairing of salumi with melon is a heavenly one. But in winter, melon is not in season and citrus fruits are, so that's what we pair with salumi. Grapefruit's bittersweet acidity goes perfectly with the salty, fatty coppa and peppery arugula, making this a great variation of a classic.

**8 ounces/225 grams coppa, sliced paper-thin**
**2 cups/56 grams arugula, cleaned**
**¼ cup/85 grams red onion, sliced paper-thin**
**20 grapefruit segments**
**¼ cup/65 milliliters extra virgin olive oil**
**Salt and freshly ground black pepper**

**1.** Arrange the coppa on four plates.

**2.** Combine the arugula, onion, and grapefruit in a bowl. Toss with the olive oil and salt and pepper to taste. Arrange on top of the coppa and serve.

**Serves 4**

## Salumi with Shaved Parmesan and Watercress

Spicy, peppery watercress with salumi is another fabulous pairing. It acts almost like a seasoning for the meat.

**8 ounces/225 grams prosciutto, coppa, or speck, sliced thin**
**2 cups/56 grams watercress, cleaned**
**¼ cup/65 milliliters extra virgin olive oil**
**Salt and freshly ground black pepper**
**2 ounces/56 grams Parmigiano-Reggiano cheese, shaved or sliced into curls with a vegetable peeler (about ½ cup)**

**1.** Arrange the sliced meat on four plates.

**2.** Put the watercress in a bowl and toss with the olive oil and salt and pepper to taste. Arrange the salad on the sliced meat, top with Parmesan, and serve.

**Serves 4**

# Grilled Figs with Prosciutto

Figs are delicious as they are, but with some heat to intensify the flavors and a little char from the grill, they become exquisite. Their sweetness, with the slight bitterness of the char, marries beautifully with the savory saltiness of prosciutto.

**4 ripe figs**
**2 tablespoons/32 milliliters extra virgin olive oil**
**Salt and freshly ground black pepper**
**8 thin slices (4 ounces/110 grams) prosciutto**

**1.** Build or start a medium-hot fire in a grill.

**2.** Cut the figs lengthwise in half. Put in a bowl and toss with the olive oil and salt and pepper to taste.

**3.** When the fire is ready put the figs cut side down on the grill. Grill, turning once, just long enough to sear them on both sides with grill marks.

**4.** Wrap each fig half in a slice of prosciutto, arrange on plates, and serve.

**Serves 4**

## Figs, Prosciutto, and Gorgonzola

Adding a blue cheese funkiness to the fig and prosciutto combo enhances both the meat and the fruit. We've included walnuts for crunch and flavor.

**8 ounces/225 grams prosciutto, sliced paper-thin**
**4 ripe figs, quartered**
**4 ounces/110 grams Gorgonzola cheese, crumbled**
**½ cup/56 grams walnut halves, toasted**

Arrange the prosciutto on four plates. Arrange 4 fig quarters on each plate, top with the Gorgonzola and walnuts, and serve.

**Serves 4**

## Cotechino with Lentils

A big fat sausage—here cotechino, which is loaded with gelatinous skin—braised with lentils is a classic Italian preparation and a great fall and winter dish. The sausage must be very well cooked to tenderize the skin used to make it.

**THE COTECHINO**
**2 pounds/1000 grams cotechino**
**½ cup/120 grams finely diced onion**
**¼ cup/60 grams finely diced leeks**
**1 garlic clove, chopped**
**¼ cup/65 milliliters extra virgin olive oil**
**¼ cup/60 grams finely diced carrot**
**¼ cup/60 grams finely diced celery**

*continued on next page*

2 cups/480 milliliters pork or chicken stock (page 244)
Salt and freshly ground black pepper
Cayenne pepper

THE LENTILS
1 cup/225 grams green lentils, rinsed and picked over
1 medium onion, cut in half
1 large carrot, cut lengthwise into quarters
3 garlic cloves
1 large bay leaf
4 cups/1 liter water
¼ cup/60 milliliters extra virgin olive oil
¼ cup/60 milliliters red wine vinegar
Salt and freshly ground black pepper

**1.** For the cotechino, prick the sausage several times with a needle. Put it in a pot, cover with cold water, and bring to a boil, then reduce the heat and simmer for 1½ hours, or until the sausage is tender. Drain and set aside to cool.

**2.** For the lentils, combine the lentils, onion, carrot, garlic, bay leaf, and water in a medium saucepan, bring to a gentle boil, and cook until the lentils are tender; about 20 minutes.

**3.** Meanwhile, sauté the onion, leeks, and garlic in the olive oil in a skillet until softened, a few minutes. Add the carrot and celery and cook until the vegetables begin to soften, 5 minutes. Add the stock and bring to a simmer.

**4.** When the lentils are done, drain them and discard the onion, carrot, garlic, and bay leaf. Put the lentils in a large serving bowl, stir in the olive oil and vinegar, and season with salt and pepper to taste. Set aside.

**5.** Slice the sausage into ¼-inch/6-millimeter rounds. Add to the vegetables and reheat.

**6.** Spoon the cotechino and vegetables over the lentils and serve.

**Serves 6**

# Bollito Misto

Inspired by our last meal in Italy at a lovely little restaurant called La Buca, in the town of Zibello in the heart of Emilia-Romagna, Brian created this classic dish of boiled meats, much as we had it there. The elderly proprietess came to our table, made a few inquiries, and then told us what we would be having, which was fine by us. It began with a large plate of her own culatello, thinly sliced, which she served with bread and curls of butter. We shared the bollito misto, which was served with a variety of condiments, an essential part of this dish. Brian created a vivid green, herby salsa verde and a red pepper aioli, as well as the star of the dish, mostarda, a sweet hot chutney. His recipe calls for a half hog's head, but you could substitute 2 pounds of skin-on pork shoulder.

**8 ounces/225 grams carrots, cut into medium dice**
**4 ounces/110 grams celery, cut into medium dice**
**2 medium onions, stuck with 2 whole cloves each**
**1 bouquet garni—½ leek, 1 celery stalk, 1 bay leaf, thyme sprig, and parsley sprig, tied up with kitchen string**
**One 2-pound/1000-gram lean beef shoulder roast**
**2 pounds/1000 grams breast of veal**
**½ large hog's head (brains removed)**
**2 large ham hocks**
**1.5 pounds/675 grams beef tongue**
**6 quarts/5.67 liters chicken stock (page 244), or as needed**
**One 12-ounce/335 gram cotechino**
**One 4-pound/2000-gram chicken**
**Salt and freshly ground black pepper**

**FOR SERVING**
**Salsa Verde (recipe follows)**
**Red Pepper Aioli (recipe follows)**
**Mostarda di Cremona (recipe follows)**

*continued on next page*

**1.** Combine the vegetables, bouquet garni, beef shoulder, veal, hog's head, ham hocks, and tongue in a large stockpot and add the chicken stock. The liquid should cover the ingredients by 2 inches/5 centimeters. Cover and bring to a boil. Remove the lid, reduce the heat to a simmer, and cook for 2 hours, skimming the foam off the top on a regular basis. (Be sure to maintain a gentle simmer, just light bubbles coming to the surface, to ensure the broth will be clear, not cloudy.)

**2.** Meanwhile, prick the cotechino in several places with a needle or the tip of a sharp knife. Put it in a pot, cover with water, and simmer until tender, about 30 minutes. Drain and set aside.

**3.** After 2 hours, add the chicken to the stockpot, along with the cotechino, and simmer for another 1 hour. Remove from the heat. Carefully transfer the meats and chicken to a platter, season the broth with salt and pepper, and set the broth aside.

**4.** When the meats are cool enough to handle, slice the beef shoulder. Remove any heavy fat and sinew from the veal and cut into chunks. Pick the meat from the hog's head and ham hocks. Peel and slice the tongue. Cut the chicken into serving pieces.

**5.** To serve, arrange the meats on a deep serving platter. Ladle some of the broth over, and serve with the condiments.

**Serves 8**

## Salsa Verde

**½ cup/28 grams fresh basil leaves**
**½ cup/28 grams fresh mint leaves**
**½ cup/28 grams fresh parsley leaves**
**4 large anchovy fillets**
**1 tablespoon/16 grams capers, rinsed**
**2 tablespoons/30 milliliters Dijon mustard**

**2 tablespoons/30 milliliters white wine vinegar**
**½ cup/120 milliliters extra virgin olive oil**
**Salt and freshly ground black pepper**
**Cayenne pepper**

Combine the basil, mint, and parsley in a blender, and process until smooth. Add the anchovies, capers, mustard, and vinegar and blend well. With the blender running, slowly add the oil, allowing the sauce to emulsify. Season with salt, pepper, and cayenne to taste. Store, covered, in the refrigerator for up to 1 week.

**Yield: 2 cups/450 grams**

## Red Pepper Aioli

**2 large egg yolks**
**2 tablespoons/30 milliliters freshly squeezed lemon juice**
**2 tablespoons/30 milliliters Dijon mustard**
**½ cup/100 grams julienned Marinated Roasted Red Peppers (page 210)**
**½ cup/120 milliliters extra virgin olive oil**
**½ cup/120 milliliters canola oil**
**Salt and freshly ground black pepper**
**Cayenne pepper**

Combine the yolks, lemon juice, mustard, and roasted pepper in a food processor and puree until smooth. With the motor running, slowly add both oils, allowing the sauce to emulsify. Season with salt, pepper, and cayenne to taste. Store, covered, in the refrigerator for up to 1 week.

**Yield: 2 cups/450 grams**

*continued on next page*

## Mostarda di Cremona

2 Bartlett pears (12 ounces/335 grams)
2 cups/725 milliliters water
3½ cups/665 grams sugar
2 plums (10 ounces/280 grams), halved and pitted
1 cup/110 grams bing cherries, halved and pitted
4 apricots (5 ounces/140 grams), halved and pitted
4 peaches (12 ounces/335 grams), halved, pitted, and each half quartered
10 ounces/280 grams (about 1 pint) Black Mission figs, halved or quartered, depending on size
2 cups/725 milliliters white wine vinegar
½ cup/65 grams Colman's dry mustard
1 tablespoon/12 grams mustard seeds

**1.** Peel the pears, cut lengthwise in half, and remove the cores. Cut each half into 8 even wedges, then cut the wedges in half.

**2.** Bring the water to a boil in a medium pot. Slowly add the sugar and bring back to a boil to ensure the sugar is dissolved. Turn the heat down to a simmer. Add the pears and cook until they are almost soft, about 8 minutes. Add the plums and cook for 5 minutes. Add the cherries and cook for 5 minutes. Add the apricots, peaches, and figs and cook until all the fruit is tender but not too soft.

**3.** Meanwhile, bring the vinegar to a boil in a medium saucepan and dissolve the mustard in it. Add the mustard seeds and remove from the heat.

**4.** With a slotted spoon, lift the fruit out of the syrup and transfer to a storage container.

**5.** Add the vinegar to the syrup and boil until reduced to a thick syrup.

**6.** Pour the syrup over the fruit and let cool, then cover and chill before serving. This will keep for up to a week refrigerated.

**Yield: 1 quart/1 liter**

# Sautéed Pork Kidney with Wild Mushrooms, Shallots, and White Wine

If you buy a half or whole hog, often the kidney will still be attached—hopefully, surrounded by rich fat. Remove it from the carcass and pull off any heavy exterior fat (this fat is especially good to render and use in pastries). The kidney should have a nice brown color (younger animals will be a little paler), and it should be firm, dry, and plump. Pork kidneys are stronger in flavor than lamb kidneys and are best soaked in milk for at least an hour or in the refrigerator overnight before using. Although the kidneys lend themselves well to stewing, in this recipe, we call for a quick sauté to keep the kidney moist and tender. It is finished with a veal stock reduction (veal is best, but you can use chicken stock; see page 244).

**1 whole hog's kidney (about 1 pound/450 grams)**
**Milk as needed**
**Kosher salt and freshly ground black pepper**
**4 tablespoons/60 grams unsalted butter**
**2 tablespoons/72 grams finely chopped shallots**
**10 ounces/280 grams wild mushrooms, such as porcini, morels, or chanterelles, cleaned and cut into ¼-inch/6-millimeter slices**
**6 tablespoons/100 milliliters dry white wine**
**6 tablespoons/100 milliliters rich veal or chicken stock**
**½ cup/60 grams chopped fresh parsley**

1. Remove any white membranes from the outside of the kidney. Cut the kidney in half lengthwise and remove and discard the hard white core. Wash it well in cold water, then soak it in milk to cover for at least an hour or, preferably, overnight, refrigerated.

2. Drain the kidney, pat it dry with paper towels, and cut it into ¼-inch/6-millimeter slices. Season with salt and pepper.

*continued on next page*

**3.** Melt the butter in a large heavy skillet over medium-high heat. Add the shallots and sauté until soft and starting to brown, a few minutes. Add the kidney and mushroom slices to the pan and sauté for 2 minutes. Turn the kidney slices over, add the wine, and cook until the wine has nearly evaporated. Add the stock and parsley and bring to a simmer, being careful not to overcook the kidney slices (they should be medium-rare). Serve immediately.

**Serves 4 as an appetizer**

# Sources

For all sausage making and dry curing supplies—curing salts, casings, bacterial cultures, kitchen tools—we recommend Butcher & Packer Supply Company. We've been working with this company for many years now for the quality of its products and the excellence of its service.

**Butcher & Packer Supply Company**
1780 E. 14 Mile Road
Madison Heights, MI 48071
248-583-1250
www.butcher-packer.com

SCALES

We recommend the My Weigh Triton T2 300 for weighing seasonings, and the My Weigh KD-8000 or the My Weigh UltraShip 35 for weighing larger quantities of meat and fat.

**Amazon.com**

**Old Will Knott Scales**
10750 Irma Dr., Unit 4
Northglenn, CO 80233
877-761-0322
www.oldwillknottscales.com

PORK

**Niman Ranch**
1600 Harbor Bay Parkway
Suite 250
Alameda, CA 94502
866-808-0340 or 510-808-0330
www.nimanranch.com

**Heritage Foods USA**
Box 198, 402 Graham Ave.
Brooklyn, NY 11211
718-389-0985
www.heritagefoodsusa.com

**American Livestock Breeds Conservancy**
P.O. Box 477
Pittsboro, NC 27312
919-542-5704
www.albc-usa.org

**Local Harvest**
www.localharvest.org
Connect with local farmers.

BOAR

**Broken Arrow Ranch**
3296 Junction Highway
Ingram, TX 78025
800-962-4263
www.brokenarrowranch.com

**Pasture to Plate**
5105 West Ogden Ave.
Cicero, IL 60804
708-652-3663
www.pasture2plate.com

SOME OF THE TOP SALUMI MAKERS

**Benton's Smoky Mountain Country Hams**
2603 Hwy. 411 North
Madisonville, TN 37354
423-442-5003
www.bentonscountryhams2.com

**Boccalone**

***Boccalone Salumeria***
1 Ferry Building # 21
San Francisco, CA 94111
415-433-6500

***Boccalone Plant***
1924 International Blvd.
Oakland, CA 94606
510-261-8700
www.boccalone.com

**Creminelli**
310 Wright Brothers Dr.
Salt Lake City, UT 84116
801-428-1820
www.creminelli.com

**Fra' Mani**
1311 Eighth St.
Berkeley, CA 94710
510-526-7000
www.framani.com

**La Quercia**
400 Hakes Dr.
Norwalk, IA 50211
515-981-1625
http://laquercia.us

**Newsom's Hams**
208 East Main St.
Princeton, KY 42445
270-365-2482
www.newsomscountryham.com

**Olympic Provisions**
107 SE Washington St.
Portland, OR 97214
503-954-3663
www.olympicprovisions.com

**Salumeria Biellese**
378 8th Ave.
New York, NY 10001
212-736-7376
www.salumeriabiellese.com

**Salumi**
309 Third Ave. South
Seattle, WA 98104
206-621-8772
www.salumicuredmeats.com

**Oyama Sausage Company**
Box 126, 1689 Johnston St.
Grandville Island Public Market
Vancouver, BC V6H 3R9
Canada
604-327-7407
www.oyamasausage.ca

DMT SHARPENING STONE

This is not really an actual stone but rather a perforated diamond-coated metal plate that works better than any stone we've encountered. We recommend using it frequently to keep your knives sharp.

**DMT (Diamond Machining Technology)**
85 Hayes Memorial Drive
Marlborough, MA 01752 USA
800-666-4368 or 508-481-5944
www.dmtsharp.com

TRAPANI SEA SALT

**The Meadow**

***The Meadow—Portland***
3731 North Mississippi Ave.
Portland, OR 97227
503-288-4633 or 888-388-4633

***The Meadow—New York***
523 Hudson St.
New York, NY 10014
212-645-4633 or 888-388-4633
www.atthemeadow.com

**SaltWorks**
16240 Wood-Red Rd. NE
Woodinville, WA 98072
800-353-7258
www.saltworks.us

FENNEL POLLEN, PAPRIKA, AND SPICES

**Zingerman's**
422 Detroit St.
Ann Arbor, MI 48104
888-636-8162
www.zingermans.com

**The Spice House**
1512 North Wells St.
Chicago, IL 60610
312-274-0378
www.thespicehouse.com

**Sausage Debauchery**
61 Woodlands Dr.
Tuxedo Park, NY 10987
845-712-5311
www.sausagedebauchery.com

**La Tienda**
1325 Jamestown Rd.
Williamsburg, VA 23185
757-253-1925 or 888-331-4362
www.tienda.com

# Acknowledgments

Big thanks first of all to Marc Buzzio for all his help, wisdom, and generosity, and also to his business partner Paul Valetutti; there's not a better salumeria in the country than their Salumeria Biellese. We also are indebted to the world-class curemaster and chef, Paul Bertolli, of Fra' Mani Handcrafted Foods, and to chef Chris Cosentino of Boccalone, both in the Bay Area, and to Armandino Batali, of Salumi in Seattle.

Chef Jay Denham, of The Curehouse and Woodlands Pork, was instrumental in our understanding of Italian butchery, especially in the general and finer points of butchering and curing culatello.

We thank Nic Heckett, also of Woodlands Pork, for leading our explorations of salumi in Italy.

We regularly leaned on Victor and Marcella Hazan, experts in Italian cookery and culture, when Italian tradition and terminology grew confusing; they were always there to guide and clarify.

Brian would like to thank the administration and students at Schoolcraft College, where he is a chef-instructor, for their help in the creation of this book (even when they didn't know they were doing so). He is likewise grateful to his colleagues who teach with passion: Dan Hugelier, Jeff Gabriel, Joe Decker, Kevin Gawronski, Shawn Loving, and Marcus Haight. He thanks his Schoolcraft sous chef, Lisa Hysni, as well.

He thanks chef David Gilbert for running the restaurants while Brian was working on this book (and Michael remains grateful to David for saving his ass in American Bounty all those years ago).

We would like to thank our assistants Emilia Juocys, whose enthusiasm and organization skills are unsurpassed, and the excellent Zach Kucynski.

Thanks also to Donna Turner Ruhlman who photographed the butchery sessions from which the illustrations were created.

We thank Maria Guarnaschelli, who encouraged us to continue our exploration of meat curing, and her assistant Melanie Tortoroli.

Judith Sutton, once again, did her usual heroic work in copyediting this book.

Neither of us would be who we are or be able to do what we do without our wives, Julia and Donna, and our kids, whom we cherish even when they don't cherish us.

Finally we bow to those farmers and caretakers of the hog who do it well, who raise the animals by hand and oversee their humane treatment and humane slaughter, as we bow to the noble beast itself, the great and glorious hog.

# Index

Note: Page numbers in **boldface** refer to recipes themselves; page numbers in *italics* refer to illustrations.

R